The End of Medicine

The End of Medicine

HOW SILICON VALLEY
(AND NAKED MICE)
WILL REBOOT YOUR DOCTOR

Andy Kessler

Collins
An Imprint of HarperCollinsPublishers

HarperCollins books may be purchased for educational, business, or sales
promotional use. For information, please write to: Special Markets
Department, HarperCollins Publishers, 10 East 53rd Street,
New York, NY 10022.

Designed by William Ruoto

Library of Congress Cataloging-In-Publication has been applied for.

ISBN-10: 0-06-113029-X
ISBN-13: 978-0-06-113029-8

06 07 08 09 10 DIX/RRD 10 9 8 7 6 5 4 3

To Nancy, my life sustainer

Contents

■

Introduction

Chapter 1

■

Broken Neck?

We were on our third pitcher when the conversation started getting interesting. I used to ski just about every year with Brad Miller. I'd known the guy forever, but then again, you never really know someone until the third pitcher. He was an equity salesman at Morgan Stanley out of San Francisco, back when I was an analyst following companies like Intel. I used to meet him and a few others on the ski slopes at Snowbird and Alta in Utah to, uh, discuss tech trends, at least that's what I put on my expense reports.

As the Red Hook slid down easy, Brad started babbling about a ski trip to Sun Valley, Idaho.

"It was one of those beautiful days—sun shining—it's getting warmer as the day goes on—my legs are feeling great," he started. I pretend to enjoy other people's skiing adventures but usually just think about myself in the deep stuff.

"Uh, you somehow forgot to invite me on this trip?" I joked.

"Must have been an oversight. Anyway, this is the first day of the trip. We're hitting the bumps just before lunch. You know that feeling, you hit the trough of the mogul, the snow kicks up behind you—you feel like a fucking rock star."

"Doesn't happen to me that much." I sighed.

"Then I catch my skis on some soft snow. Instead of the usual ass-over-teakettle fall, I get twisted in some weird half turn, my upper body goes 180 degrees and I end up looking uphill, my ass hits the top of the mogul and a shock wave rips up my spine—my glasses go flying, the whole damn thing."

"Jeez." Okay, now I'm paying attention.

"So I'm seeing stars—I'm dizzy, woozy, nauseous, sweating. It took me 15 to 20 minutes to get a grip. I side-track out and get on some green run and meet my buddies for lunch."

"They didn't wait for you?" I asked.

"Nah. So I immediately went for a beer, figured that should help, I'll be fine. After two sips, my neck is stiffening, seizing up, so I say, 'Guys, I'm fucked. I gotta get out of here.' And they started laughing, 'You wanna ride in the toboggan behind the ski patrol guy?' Yeah. Really funny. No fucking way."

"That would have been a sight!" I said.

"Yeah, maybe. So I rode the ski lift down, got to the bottom, and drove myself to the Wood River Memorial Clinic."

"And they patched you up?"

"I was third in line. Two people ahead of me had hit trees that day. Just my luck. I end up waiting four hours and then finally get an X ray, then a CAT scan, then another X ray. This very nice female doctor tells me the little wing piece on one of my vertebrae looks like it may have broken off. If it's floating around, it can cut your spinal cord and kill you. I needed an MRI, which they have in Boise. So I get strapped to this plastic backboard, head immobilized, and they put me in an ambulance for a three-hour ride to Boise—the long way, which had less bumps."

"Ouch."

"Just as the ambulance is about to leave, she says to me, 'Oh, yeah, there's this other thing. You have this tumor—we don't know what to think. But that's not your immediate problem, we have to deal with your neck first.' The next three hours were not the most pleasant, as you can imagine."

"What was it?"

"Well, I get wheeled into the emergency room in Boise, still strapped to this board. I waited another three hours—there were only these curtains—a little girl next to me had drunk too much Jack Daniel's and was detoxing and some old guy on the other side was puking in a bucket."

I started laughing. "Sorry, that's not really funny." I put my beer down.

"It is now. So around three in the morning, I finally get an MRI. This really cool doctor comes out and tells me, 'You didn't break your neck. Nothing to worry about there.' I felt relieved. But then he says, "However, you do have this tumor the size of a martini olive sitting on the top of your spine.' He knew a doctor at Stanford he would put me in touch with, and then he discharged me." Brad sighed. "Then the nightmare started."

"Oh, jeez."

"Yeah, you don't want go through the rest of this. The good news is that I kicked it."

"Kicked what?" I asked.

"Well, I had this tumor on my clivus, at the base of my skull. It's inoperable, too much in the way. It turns out that I had multiple myeloma—bone marrow cancer. It's one of those autoimmune things, your body generates lots of white blood cells, which basically eat your bones away from the inside. In the 1800s, people would sneeze and break ribs, that kind of stuff. It's rare—and misdiagnosed all the time—because other things can kick up your white blood cell count."

"So you had to almost break your neck skiing to find it?" I asked.

"Yeah, it's just a fluke, but luckily, it happened early enough."

"Chemo, I assume?"

"First thing, I had to find a doctor. Stanford led to USF, which did some focus radiation on the tumor to try to stop its growth. But bone marrow is messy stuff. Dana Farber at Harvard is the leading clinician in multiple myeloma. But they pointed me to M. D. Anderson in Houston, Fred Hutchinson in Seattle, which was full, and ACRC, the Arkansas Cancer Research Center in Little Rock. So off to Little Rock, every four to six weeks, 15 months start to finish."

"For what?" I asked.

"Four rounds of VAD cocktail, two rounds of DT-PACE, which has thalidomide—"

"No plans on getting pregnant any time soon?" I kidded.

"—and then two rounds of stem cell transplants. They basically pump your blood through a machine, which pulls out all of your stem cells, like 50 million of them. Then they give you this high dose of melphan, which is like battery acid. It kills all your bone marrow and then they bring you back by reintroducing your own stem cells, which regrows the bone marrow."

"I had no idea," I said in wonderment.

"It's gone. I get tested every month. $900 in blood tests—how that's paid for is another nightmare, a keg's worth of discussion. It worked—I'm sitting here drinking beers today."

"Thank God."

There was a long pause. I wasn't sure what to say. Sorry? How inspiring? You're a model of strength? I hope to hell that doesn't happen to me?

"So what you're telling me is that skiing saved your life?" I asked.

"You could put it that way," Brad replied.

"I'm going to use that line with my wife." I laughed.

Brad laughed, too. I stared at him and just shook my head. The whole thing was unbelievable. I felt a twinge in my neck. I unconsciously rubbed my right hand on the top of my spine and thought I felt something mushy. Brad started looking at me funny and I moved my hand and scratched my head and chugged the rest of my beer. Shit. I think I am going to schedule a visit to my doctor—do I even have a doctor?— tomorrow goddamn morning.

Chapter 2

■

Physical

"How long has it been since your last physical?"

"Uh, hmm, well, I don't know, probably ten years or so. How long has it been since 1989?"

"More like 15 years."

"That's about right. I had one of those mandatory physicals when I started at Morgan Stanley back then. No, check that, I thought about it, but I think I skipped it."

"Pretty healthy, then?"

"As far as I know."

What I meant to say was that I'd been sitting in just my white BVD u-trow on a raised bench covered with a flimsy piece of paper for, let's see, 45 minutes, waiting for Dr. Welby or Kildare or whatever his name was. I'd probably caught a cold or virus or infection or something from the last patient.

"Then why are you here?"

"40-year physical." Couldn't quite say my wife made me.

"Not good at math, then?"

"What?"

"Says here you're 45."

"I've been busy."

"Mmm-hmm. Okay, let's get started. Pulse fine, blood

pressure a tad high. Weight . . . well, a little high, maybe 15 pounds. Recent weight gain?"

"It's been the same for probably 15 years."

"Smoke?"

"No. Does a cigar once a year count?"

"Drink?"

"Constantly."

"Drugs?"

"Not since Woodstock." Deadpan face. Not even a crinkling of the corners of his mouth.

"Red meat?"

"Preferably dripping."

"Okay, say aah." He reached for a wooden stick and jammed my tongue down hard.

A stethoscope front and back. A look in both ears. The old "flashlight in the eyes" thing. And then out came the rubber hammer, tapping my knee. I almost kicked the doctor where it hurts, but then remembered about paybacks and he hadn't yet done the "got you by the short hairs" thing.

"Okay, cough for me . . . good, good."

Now I could kick him. A rubber hammer? Jeez. This routine hadn't changed much since my visits to my pediatrician. Has medicine gone sideways in the last 30 years?

I didn't have much time to ponder this when I noticed my doctor had a slight grimace on his face as he pulled a rubber glove on his right hand.

"Okay, bend over . . ."

("And say yee-hah?") I wasn't sure if I actually whispered or just thought these words but it was over before I knew it.

"Okay, you're fine. Good, well. Everything's fine."

"Great. I'll be back in another ten years . . ."

"Well, we have a little more to do, let me go over your blood tests."

Right. I had forgotten that I had already been to the clinic last week when a giant nurse vampire sucked what seemed like a gallon of blood out of my vein.

"Okay."

"It looks pretty normal. Glucose, creatinine, albumin, globulin. Your PSA is 1.42, we like to see it under 4." Whew. "Well, except for your cholesterol. Overall, it's on the high side. Your so-called good cholesterol, HDL, is 30. Under 35 is low. Let's see, your bad cholesterol, LDL, is way high, 162. Over 159 is borderline high risk. We usually like to see it the other way around."

No kidding.

"What does that mean?" I asked.

"It means you need to lower your cholesterol."

"Or I'll have a heart attack? Stroke? Something like that?"

"Could be. But basically, we like to see lower cholesterol. You eat eggs?"

"Sure, but—" I said.

"And you told me you eat a lot of red meat."

"Some, not every day."

"Well, I would try salmon, lots of omega 3s . . ."

"I have salmon when I'm in Seattle—"

"Chicken's okay, no skin. There is pork, the other white meat."

"I've seen the commercials."

"I'd work on getting that LDL number down. You're still young, but if it keeps up, we'll probably have to prescribe statins to get the number down."

"Am I at some sort of risk?"

"It's not really clear. The literature suggests it is healthier to have a lower LDL number, and I tend to agree."

"But why?"

"Oh, yeah, do you exercise?"

"I play basketball. Is there something I need to worry about?"

"Sure, your cholesterol number. Okay, thanks. Let's reschedule another visit next year so we can track this." And with that, he was out the door and I was left in my underwear sitting in a puddle of sweat.

I walked out to the parking lot, turned to look at the building and did what I should have done inside—I flipped him the *digitus in medio*. C'mon, the guy didn't tell me anything. "Borderline high risk." "Literature suggests." Gee, thanks for nothing, pal. Do I have a throbbing tumor on my neck? Are my arteries clogged up like the bathrooms at an old-age home after chili night? I could check my credit report, run diagnostics on my car, ping a router in Seattle in under a quarter of a second, but I had no clue whatsogoddamnever if I was healthy or not. And all my seasoned professional doctor with Ivy-tainted sheepskins on the wall could do was hit me in the knee with a rubber fucking hammer? He might as well have started chanting and dancing barefoot.

Okay, I'd calmed down. I thought about it that night, over a bottle of wine, and decided that some real changes were needed in my life. I was turning over a new leaf.

"Honey, I've made a decision. I'm getting eggs out of my life. Don't they have those Egg Beaters at the supermarket? From now on, it's just sausage, cheese, and Egg Beater omelets for me."

It had taken almost three months to get an appointment for my physical, but not two weeks later I got a bill from my doctor. $408 for my "Preventive Visit." No wonder I had been preventing it for 20-odd years. All for a measly eight minutes of his time and a lecture on cholesterol I could have gotten for $6.95 from *Men's Health* magazine. Plus, there was another $32 added on from the lab that took my blood, $16 for the "Collection of Venous Blood" and another $16 for "Specimen Handling." What a racket.

The next day another bill arrived from the outside blood lab. Another 220 bucks. Did they plate my blood in gold? Inject it in laboratory mice? Did a village of pygmies walk through it barefoot to discern its contents? Maybe someone gargled with it and noted its various tastes?

When the time comes to see if those Egg Beaters were worth a damn, no way was I spending another $600 on Dr. Greedy. So I marched into the medical records department and shelled out $15 in copying fees to obtain my very own copy of my personal medical records.

I thought they could just hit print. No way. Instead, I got pages and pages of test results, a couple of check marks on meaningless forms and the doctor's write-up claiming I was a well-developed, well-appearing gentleman in no acute distress. As my grandfather used to say, that and a nickel won't get you on the subway.

No matter. From here on, I was going to be my own doctor. And why not? My kids hit me in the knee with a real hammer every couple of weeks. I had a home blood pressure machine. Egg Beaters weren't half bad, and surely I could read the results of a blood test and look up my cholesterol number to see if it was high or low.

Part I

Chapter 3

■

In the Cath Lab

Yech. Nasty." A cardiologist was standing beside me look-ing over my body at the screen. "What a mess. You see it?" I looked over and her face was contorted like a shot of pure lemon juice had just hit her tongue. I, on the other hand, was starting to feel dizzy. Large spurts of deep red blood had just arced a few feet skyward and then back down to the table, quickly soaking a blue-green pad until the doctor capped the catheter with a syringe filled with clear liquid. "See the old stents? It's stent city in there," I heard her say.

Not to worry. That wasn't me on the operating table. My cholesterol wasn't that bad, or at least I hoped it wasn't. I was sitting on the observation side of a lead-lined glass window and alternated glances between the patient on the table and an LCD screen showing his arteries. I felt like I was in a recording stu-dio. Wearing a white jacket, I could easily have been mistaken for a doctor, except for two small details: I had a "what the fuck" look of amazement on my face and a tag hung from an elastic band around my neck that read "Visiting Student." I needed to get up to speed on medicine and what goes on in hos-pitals, so I'd signed up for a week-long program at a big New York hospital that promised seminars plus the chance to follow doctors around.

• • •

An hour earlier that morning, I had been assigned to Dr. Kam
(not his real name, nor are his patients' names real), a third-
year fellow in the cardiology unit. Dr. Kam was probably 29 or
30, with shoulder-length signature Asian jet-black hair parted
down the middle. It was eight o'clock in the morning and he
was already cooking, only slowing down long enough to say,
"Okay, come on, then," to me and an 18-year-old college freshman named Troy.

We started strolling through the maze of halls in the hospital. I was completely lost. I thought I heard "Dr. Howard, Dr.
Fine, Dr. Howard" over the PA system.

"So you guys are considering becoming doctors?" Dr.
Kam asked as he fixed his hair net.

"Yes, sir," piped in Troy.

"Well, it's not too late to change your . . ." His beeper
started going off. He had a pissed-off look until he read the
numbers. "Okay, we got one, let's go."

We followed him, struggling to keep up with his pace,
dodging a few gurneys with swinging IV drips. We came to a
screeching halt in front of a door that read "Cath Lab 4."

"Wait here," we were instructed.

I made idle chitchat with Troy as an army of scrubs scurried in every direction.

"Are you taking Chem 207?" I asked.

"Yes, sir. It's challenging, sir." I suppose he had already
figured out, by my middle-aged hangdog look, that I had taken
the course 25 years before, but he looked too nervous to ask.

Dr. Kam popped out of the heavy door, pulling on what
looked like a bulletproof vest. "Okay, we've got a live one." He
looked excited. "Just sit in here and be quiet."

I nodded while Troy answered for us with a "yes, sir."

Lying on the table was a very awake 70-year-old named
Jorge something or other. Dr. Kam was pulling up his records

on a rather cheap-looking Dell PC. Looking over his shoulder, besides Troy and me, was another 30-year-old in scrubs, Dr. Prakash, an articulate Pakistani fellow, I guessed.

"Looks like he was here in March, pull up the Cardio-Logica," Dr. Prakash said. Dr. Kam clicked away until a screen that looked like a computerized version of the game Operation, seemingly with the big red nose and all, came on the screen.

There were lots of Harumphs and Hmms and Shits and No Ways until Dr. Kam stood up and said, "You two, we're going to the reading room." Neither Troy nor I knew what to do, so we followed them, again struggling to keep pace. We followed them into a room filled with screens and CDs and a large Sony Optical Drive jukebox I recognized. It could have been ABC's horse trailer from *Monday Night Football*. Dr. Kam typed in Jorge's client number, clicked on an icon labeled March and then started flicking a giant shuttle knob that I recognized from my VCR and that was fast-forwarding and freezing a black-and-white but mostly gray video on a screen. Every few seconds, it would flash brighter and even I, a biology neophyte, could make out arteries, but with a whole bunch of inch-long mesh tubes.

Kam and Prakash were pointing and muttering back and forth, and every once in a while remembering we were there, would ask, "You see the stents?"

I answered, "Yes, sir." It was catchy.

Troy asked, "So this is his last . . ."

"Yup, from nine months ago. We want to see what they did before we go in there. We're just playing it back."

"Wow, in case he has a . . ."

"He had an AMI."

"A what?" Troy asked.

"An acute myocardial infarction."

"Wha?" Troy mumbled.

"He had a heart attack a few hours ago, we're going in."

For duty and humanity. Dr. Howard, Dr. Fine . . .

•   •   •

What I hadn't realized was that this week-long seminar was for college kids, premeds. Mostly young women in pointy shoes. I should have remembered to read the fine print. No matter, it was eye opening. One of the reasons I had signed up was that you got to shadow doctors, mostly cardiologists it turns out, around the hospital.

Dr. Kam's bulletproof vest was actually a lead jacket. Covering chests, genitals and throat. It didn't look comfortable. The woman cardiologist with the puckered face was very pregnant looking, probably why she was observing along with Troy and me.

"It's pretty simple," she explained. "They open a hole in the groin area and push a catheter up the main artery toward the heart. It's a pretty direct trip. That wire you see is curled just right so that it ends up in the coronary arteries. Every once in a while, they'll squirt some dye in there and it will light the arteries up. That's this screen." She pointed.

A third doctor, who looked like the Irish bartender from the restaurant I'd eaten at the night before, was pushing what looked like a fiber-optic cable into Jorge. A nurse practitioner (I saw the NP on her badge) was in a lead-lined outfit in the room, and a registered nurse (RN, duh) was in the observation room with us, monitoring an EKG and fetching equipment as needed from the hallway. Dr. Kam said in a soothing voice, "Okay, you're going to feel a warm flash in your chest, that's just us taking a look." I hadn't passed out yet, but I was leaning pretty hard against the wall to hold myself up. There was no more spurting blood, which helped.

"Those mesh tubes are the old stents. You can see pretty clearly that his arteries are clogged between a few of the old ones, and even in this other fork." Again she pointed.

No shit. It looked like an accident on the Brooklyn-Queens Expressway at 5:15 in the evening. Five lanes collapsed to one.

"It's a standard 526."

I had read about angioplasty and stents and the promise of drug-eluting stents. Jorge was soon to be the proud owner of three new ones. I was mesmerized watching it take place in front of me. I couldn't quite make out the balloon blowing up the artery but definitely could see the inch-long stents being moved back and forth until they were in just the right spot. I had been to Disneyland a few weeks before with the kids, but this was the E ticket ride. I looked over and Troy was stifling a yawn. I made a mental note to have a salad for lunch. Dr. Kam finished up and motioned for us to wait for him out in the hall.

Thank god for Smartphones. College must be so much easier today. I don't know about you, but instead of nodding my head and pretending I knew what a professor was talking about, now I could look it up and not look and feel so stupid.

Out in the hall, I plugged in 526 and Medicare and a few other search terms and out popped a few startling facts.

526 means Diagnostic Related Group 526, our government's coding system for medical care that's been around since 1983. You know you are in trouble when something is created by the Social Security Amendments of 1983, Law HR-1900 (PL 98-21), Prospective Payment for Medicare Inpatient Hospital Services.

Dr. Kam's Percutaneous Cardiovascular Procedure w/Drug-Eluting Stent with AMI (after that heart attack thing) is, according to some bureaucrat, worth $14,784.

Whether it costs that much is someone else's problem. Whether it's worth that much is, well, a matter of perspective. To Jorge, it's priceless. To Medicare, which pays for all of these things if you are over 65, it's damn expensive. Not only did I get to watch Jorge's procedure, but in some small way, I paid for it and so did you.

As far as I could tell, health care in America is like Saturn—it's incomprehensibly big and made out of lots of moving parts, yet far enough out of our reach that we can't truly understand how massive it really is. Depending who you ask, around $1.8 trillion was spent on health care in 2005. That's "t" as in trillion. Carl Sagan talked about billions and billions of stars. That's bubkes. With the U.S. economy at about $12 trillion, that's 15% of the entire economy spent on doctors, hospitals and aspirin. Is that big, small, just right? Will technology help or hurt health care? Does it just get so expensive it gets rationed? Will some miracle breakthrough get stillborn for lack of funding?

No one quite knows. But for some odd reason, I'm standing with a white coat on in the hall of some smelly hospital with lots of sick people and blood spurting all about, trying to make sense of all this.

Dr. Kam stripped off the lead vest, washed his hands, left his funny green hairnet-like hat on and was out the door in a flash, so Troy and I in our white coats were up and at 'em.

We stopped in front of the door of a two-person hospital room. "Wait here, I need to ask if it's okay if you come in."

His head poked out. "Okay, you're cleared."

"How you feeling?" Dr. Kam asked the patient, who, by the way, had a drop-dead gorgeous view of the East River.

"I feel great," he said.

His very concerned-looking wife piped in, "He seems fine."

"Does it hurt when I do this?" Dr. Kam asked as he pressed down on his groin.

"A little." (So stop doing this, ba-dum-bum . . .)

Dr. Kam didn't wait for the response. He typed his password into the PC in the room, pulled up the patient's records, vital stats from overnight.

"Okay, you can go home, everything looks clean."

I heard the patient say, "Really?" as I quickly followed Dr. Kam out the door. He paused to squirt some antiseptic liquid on his hands and rub. I did the same.

"We did a 526 on him yesterday, he's fine."

His pager went off. With another annoyed look, he checked it and started laughing.

On the way out, Dr. Kam nodded hello to another doctor passing by.

"Nice hat." His buddy smirked.

"Uh-huh," Dr. Kam nodded.

"Your parents must be very proud of you, what with your new job scooping potatoes."

We repeated the same routine: wait outside, come in, check the groin, check the vitals, "you're outta here," a few clicks on the PC, and we're onto the next one. Each time, the doctor reached over the counter of the nurse's station and pulled out a different three-ring binder, tabbed toward the back, scanned some indecipherable handwritten notes and added some of his own. He saw my confused look and said, "So much for electronic medical records, huh?"

The next time through, we stopped for a while at the nurse's station. Dr. Kam logged in again and studied the notebook, shaking his head. His pager went off again. He laughed. I gave him a what's-so-funny look and he handed me his pager. 6400619.

I didn't get it and must have looked it.

"It's the BIG DONG page," he explained.

"What?"

He turned the pager upside down.

"Oh, yeah." I sighed.

Troy was snickering.

"Doctor humor. Someone's got too much time on their hands this morning."

We went in to see another patient.

"How you doing?" Dr. Kam asked.

"I feel fine."

"Yeah, you are doing great. But we found some microscopic-size damage during tests. It probably means nothing, but we may have to keep you for another night to check a little more," Dr. Kam deadpanned.

"You're going to get some pushback from me on that."

"It's probably nothing, just microscopic . . ."

"I've gotta be in Rhode Island—" the patient insisted.

"I'm just informing you that there is a slight chance of damage. We'll know more soon."

We left, antisepticized, and then Dr. Kam opened up. "That's what we have to deal with. Yesterday this guy couldn't even stand up, had an infarction, we do a 526, clear his arteries, save his goddamn life, yet he complains that because we are concerned about this damage, about his health, he might not be able to check out of the hospital the next day. Jeez." It was only 8:45 and I was tired just looking at Dr. Kam.

His pager went off yet again. "Probably the SHE BLO page," he said, chuckling, but then started shaking his head as he sped up and led us to Cath Lab 3.

This time, it was 68-year-old Mrs. Nussbaum (Steve Martin still has her credit card) on the table. Dr. Kam suited up, a stream of blood arced, I leaned harder against the wall to stay conscious. This time, it was just Kam and Prakash. A nice Indian guy, Dr. Vishra, sat with us observing the procedure. He had to run off to a meeting, he told Dr. Kam, so he would let the two fellows do the procedure themselves. Dr. Vishra looked at Troy and me, asked who the hell we were, and then started to explain every step of the procedure in more detail. When they flashed the dye, the arteries looked fine, not just compared to poor Jorge's but clear, 90-mile-per-hour autobahns.

I watched Dr. Vishra note 50% blockage in the Cardio-

Logica/video game Operation software. "Great. Fine. Fantastic." Dr. Vishra looked back at me. He looked like he had just swallowed an elephant. I think he was looking at me to see if I realized that he had just authorized a completely unnecessary and probably ungodly expensive procedure to rule out heart problems as the cause for Mrs. Nussbaum's chest pains. "Some people just like the comfort of knowing that everything is all clear in there." I shook my head in agreement as he slipped away to tell Mr. Nussbaum that everything was fine.

That was that. We followed Dr. Kam around for a while longer and then rejoined the seminar we were attending. I asked Troy why he was here.

"My mom made me come," he told me. Do all doctors have pushy moms?

Oh, well, I think I'll still get something out of this, even if this is the closest I will ever come to being accused of being a doctor!

A few clicks in the phone and I discovered that we had just witnessed a DRG 125—Circulatory Disorder except AMI w/Card Cath and w/o Complex Diagnosis.

Medicare (and my 1040) just coughed up $5,299 and now Dr. Vishra needs to find some other explanation for Mrs. Nussbaum's chest pains. "Did you have Mexican takeout again last night, Mrs. Nussbaum?" might have been an easier procedure than a catheter up the thigh.

Of course, I'd had a steak the size of Cleveland the night before, and could visualize my own rush-hour arteries. The operating table in the cath lab was empty at the moment, and I was wondering if I could use my Visa card to pay $5,299 to have these guys take a peek inside of me.

And there is the rub with health care today.

Chapter 4

■

What the Hell Am I Doing Here?

I've got to admit—just the smell of a hospital makes me sick. What the hell was I doing walking around with a white lab coat watching blood spurt and doctors run around and looking into the sad eyes of grieving spouses? It felt like an episode of *Twilight Zone*—I'd been dropped into this parallel space with the walls collapsing around me.

Staring at a green screen watching stock prices flicker up and down, or chatting about next-generation wireless networks was more my speed. I'd spent 20 years on Wall Street analyzing and investing in whatever was the latest and greatest technology I could find. I stuck my nose anywhere I could find costs that continually dropped. Lower costs almost always mean that some new application opens up, some new industry gets invented practically overnight to leverage cheaper costs. And almost always, people get displaced. Silicon is cheaper than people.

Scale, scale, scale. Push costs down and some new business that does things cheaper and better pops up faster than a Whac-a-Mole. Scale with a capital S. It works like magic. I made a career out of riding these waves. I wondered if I could find them in medicine. It wasn't so obvious.

This is how Silicon Valley works. Every year, the cost of

every chip, gate, transistor, function drops by 30%—it halves every two years. If nothing else happened, Silicon Valley would shrink into oblivion. But elasticity kicks in—SCALE. Put simply, every time costs decline, some new application opens up to take advantage of the cheaper functionality, and you sell three times as many. Or 10 times as many.

I've seen it with PCs, cell phones, TiVo, routers, DVD players, electronic engine controllers in autos, and video games. 3-D graphics live on this learning curve, providing more and more photorealistic animations. And just like that, because of new technology entire careers disappear like trolley car conductors of old—phone operators, tellers, stock traders, librarians, postal sorters, numbers runners for bookies, draftsmen, magazine layout experts, film editors, stock traders and on and on. Were doctors next? I secretly hoped so.

Don't get me wrong. I care about health—my own, anyway. I don't smoke. I drink only on days that end with a "y." I eat sausage but only with pepper and onions to soak up the excess fat (isn't that how it works?).

It's just that the whole doctor and hospital thing gives me the creeps. I watched a show on PBS where doctors reconstructed a knee and I had nightmares for a week. I'd have made the perfect pacing dad in the waiting room, but the damn emasculation movement of the 80s meant I had to watch all four of my sons' births—even had to cut the umbilical cords. I haven't looked at my wife the same way since.

Yet here I was in this godforsaken hospital by choice, on purpose, with intent. What gives?

I'd been a casual observer of the industry for decades. Working on Wall Street, I'd listened to plenty of pitches for the latest miracle drug or medical device. But every time I looked at health care, costs were going up, not down. That was a nonstarter for me. Where's the scale? Where's the silicon? Where's

Waldo the geek? I didn't see it. Since closing a tech hedge fund that my partner and I ran, I'd been on the prowl for something, anything that scales like Silicon Valley and displaces people. That's where I felt comfortable. I had the time, I had the energy, I had the desire, I had the resources—I just had no good ideas. Alternative energy, civil aviation, China? Nope, nope, nope.

But as I started sniffing around medicine and health care, it felt somehow different. I'm not sure how. It just did. It seemed like the pace of innovation was changing. Not that you can get out a yardstick and measure anything, but it seemed that electronics and computers almost had to become as important as biology and chemistry.

Or maybe it was just a feeling of schadenfreude for doctors, the high priests of the false god of ultimate healing.

Everyone complains about the runaway cost of health care, outpacing inflation, $1,500 added cost to every GM automobile, talk of socialized medicine. Yuck. Doctors complained about their least favorite words—managed care—but it didn't matter. Costs just seemed to always go up. But if electronics and computers were involved and doctors were walking around selling pencils in tin cups . . . hmmmm. I might just have to check it out.

But then there was that age thing. Not maturity, but age. I'm 46. I have kids I'd like to see grow up. I was at the tail end of the baby boom, so already there have been lots of people of my generation dropping like flies—heart attacks, breast cancer, lung cancer. Was I next? Who knows? It seemed like a game of roulette—the odds were low, but when your number came up, it was an ambulance ride and frantic race to patch up your arteries, or maybe worse, six months of chemo and barfing and hair loss and looking like Twiggy, who a faint breeze could knock over.

Plus, and I really hate to admit this, especially living in

Silicon Valley, technology was getting a little dull. I go to conferences about wikis and Wi-Fi, podcasts and blogs, and I always leave with an empty feeling, bored to tears. It's all great stuff, but technology somehow seems gripped with incrementalism. It's all really neat and cool and wow, but somehow predictable. Gee, in five years we'll have cheap terabyte drives so that we can, what, watch *Simpsons* reruns and shop more efficiently?

Forget that. It's all about taking control. One by one, industries are being democratized. Power is shifting from producers and service providers to users. We manage our own money, we control our own banking, we pick our own wireless plan, we comparison-shop online, we buy songs instead of albums, we print our own photos. Power to the people—everywhere except medicine. In a small way, I had decided to take control from my overcharging, wham-bam-eight-minute-consulting, rubber-hammer-wielding doctor. Maybe with the right tools, we'll all take control.

I've trained myself to look for quantum leaps, entire industries that are created overnight from some new technology— purple people-eaters. I just wasn't finding it in computers or cell phones or even consumer electronics. I have this almost chemically induced craving for big change—for quantumentalism. It just had to be out there. I'd look under rocks if I needed to.

So curiosity and opportunity got the better of me. I had to know if there was something coming down the pike to save my ass (and other body parts, too). If I could find it, find where health care scales, find some new market that doesn't just enhance my life like search engines or HDTV or Smartphones, but actually saves it—makes me live longer and healthier— man, this might just be the biggest thing going.

If technology was going to somehow fix health care, I

wanted in—as a customer, observer, investor, whatever. It certainly was a monster market. If it meant that I had to watch a little blood fly around, see organs twitching, feel faint all the time, become a hypochondriac and not sleep well for a year or so—well, this was a small price to pay.

Chapter 5

■

Echo/Nuclear

Dr. Blake put up the first slide, which read "Health Care Industry Overview." I heard a group groan and watched as half the kids, almost one by one, coughed for 10 seconds, got up, and sheepishly walked out as if they were going to the hallway to finish clearing their throats or get a drink of water. I remember inventing that move 25 years ago.

Meanwhile, I was furiously taking notes. $1.8 trillion in spending, 15% of GDP. About a third for hospital expenses, another third for doctors and clinical services, and then another quarter for drugs and 10% for nursing home and in-home care.

"And who pays?" Dr. Blake asked. "Individuals reaching into their wallets come to only 16% of health care spending. That's it. Private health insurance and other private funds pay for 40%. That leaves the government to pick up the remaining 44%. For those of you keeping track, it's 33% feds and 11% states.

"And the U.S. is tops at spending, over $5,400 in spending for every man, woman and child in America. No one else around the world is even close. The Swiss drop $3,300, Germans $2,800 and our frozen Canadian friends to the north $2,700. And the Brits, barely $2,000 per capita. Give us your tired, your poor, your huddled masses, and those with urinary tract infections." I chuckled, but no one else did.

Dr. Blake was facing a shrinking audience—20-year-old premeds with pointy shoes just didn't like all this talk about money.

"And who do we have to thank for this? Anyone, anyone? L . . . anyone? L . . . B . . . anyone?" He was imitating Ben Stein from *Ferris Bueller's Day Off* and a few kids sat up smiling and started paying attention.

"Okay, LBJ, remember him, followed Kennedy, the Great Society?" A lot of blank faces. I looked down and started laughing. "Well, anyway, Lyndon Baines Johnson was the ultimate politician. He figured he would be buying senior votes back in 1965 by offering free hospital stays to everyone over 65 and implemented Medicare, technically an insurance plan. Then came Medicaid to help the poor pay doctors' bills. As Americans, we can be bought but we ain't cheap."

He was on a roll. "And boy, do we take advantage of it, to the tune of almost $500 billion. Jane Fonda got us jazzercising, but Americans are still getting sick. 70% of that $1.8 trillion is spent on chronic disease. Something like $210 billion was spent last year on cardio and stroke, $192 billion on cancer, $75 billion on obesity-related illnesses, $92 billion on diabetes and $22 billion on arthritis. Add a couple hundred billion for pharmaceuticals and you can see how we get to that 15% of GDP.

"Is it worth it? Anyone? Anyone?" Dr. Blake had a mischievous smile. "It depends who you ask. Life expectancy was 37 years in 1850. By the late 1930s it was 60, although that was mainly from rising infant survival rates. Today it's around 77. But if you are already 65 and on Medicare, naturally, you can expect to live well into your 80s, and I can guarantee that you think the $1.8 trillion is well spent. If you are, pardon the expression, a 22- to 64-year-old working stiff, with shrinking take-home pay, you may have a different opinion."

He paused and looked around. "Now, let me tell you how to run a 10-doctor practice for fun and profit."

I coughed for 10 seconds, then got up and left.

Dr. Blake was right. Close presidential elections in Florida were a boon to health care providers, keeping federal health care spending as the third, second and probably first rail of politics.

Life expectancy is key. Some think life expectancy is now increasing one additional year for every four years that pass. Time waits for no one, but I suppose it's not in a big hurry, either.

So the trillion-dollar question is: Does health care spending over the next decade go from $1.8 trillion to $2 trillion, or do new procedures, miracle drugs and octogenarian mating rituals pop it to $3 trillion?

Can we continue to have our excellent health care system? Can we afford it? Can anyone and his brother have a DRG 125, a nice little peek into the arteries, and have you and me pick up the $5,299 tab?

Or is there a better way? Surely the techies in this country are inventing new systems to peer into our bodies and look around without catheters up the groin, which, from what I can gather, beats the alternative of cracking the sternum open. What if some new stuff comes along to detect and eradicate cardiac problems, and we can't afford it? Wouldn't that suck? How much of that $210 billion spent on cardio and stroke is really necessary? Is there waste in the system? Why was I asking so many questions when I hadn't even the simplest clue about how this stuff worked? Must be the white coat.

"What does that look like to you?" asked Dr. Bernstein, who appeared to be in charge. I was back with doctors out in the smelly hospital again.

Four of us visiting students crowded into a tiny reading

room labeled Nuclear Imaging. One, Ali, was a senior headed to med school in the fall. I asked her if she took Chem 207, and I got back a why-are-you-bothering-me look. I used to get that in college, too. Ah, you can go home.

Dr. Bernstein was in the middle of grilling three residents. The room was littered with files, books and a month's worth of *New York Post* newspapers. What looked like a five-year-out-of-date Sun workstation was on the desk; its 21-inch display had a bunch of splotchy yellow and blue and purple colored rings and U-shaped things in several rows.

"Well, it appears to show some smattering of a hint of something material . . . uh . . . can I see the vintages?" stammered Peja in a Slavic accent.

"Why do you want to bother with vintages? You sure?" Dr. Bernstein looked annoyed.

As Peja stumbled through his grilling, another fellow named Steve leaned over and whispered to Ali and me, "Those are time-sequence images of a stressed heart. The yellow is heart muscle containing the radioisotope. Rings are a side view and the Us are more lateral."

"Thanks," I whispered back.

"Normal," Peja declared.

"Yeah, that's pretty obvious. Vintages . . . jeez," Dr. Bernstein seemed to have honed his humiliation skills. "Who's next? Vinny?"

Vinny looked about 27 or so, and his side-cropped "hairdo du jour" was straight out of Brooklyn. Meanwhile, the room kept filling up. Two more residents stopped in to check out what was going on. Then a technician came in and started reading a *Post* while eating her lunch in the corner.

"It's like the stateroom scene from that Marx Brothers movie," I whispered to Ali.

"The what?" she asked.

"Never mind," I said.

"Let's go," Vinny said, rubbing his hands.

"Okay." Dr. Bernstein shuffled through his files. "This one. Jack the Lineman." He had a devilish look on his face.

"The football guy. Okay. Okay, hit me." I watched a bead of sweat drip down Vinny's long sideburns. "Let's see, left ventricle decent, lower aortic a little sparse, um, shading, um, well, uh, can I see the vintages?"

Dr. Bernstein snorted. "Remember, he's an athlete, lots of muscle between us and his heart, could it be that . . ."

Resident Steve pulled Ali and me out of the room, "Here, come with me. I'm about to do a stress test, someone's going to explode in there."

We followed him out of the stateroom and then behind a drawn curtain. A pleasant-looking 70-something-year-old man with an intravenous in one arm and a blood pressure monitor on the other was standing in front of a treadmill.

"Okay," Steve said, "we're going to start you off slow, and then increase the rate and the incline until we reach stress," Steve told him.

"Okay. I should say I haven't slept well for the last several nights."

Steve waved him onto the treadmill. "I'll be monitoring you. Hop on." Steve had a calming tone.

The man started walking on the treadmill, and every three minutes, the incline got steeper and the pace quickened, and then in a few more minutes, it got steeper and the man was walking faster and faster and on and on. I got tired just watching the poor man racewalk uphill.

Then he started huffing and puffing, really hard, but had enough breath to lash out. "You know, I had some discomfort. Dr. Tyler figured imaging might help, but he didn't tell me I was going to be running up Mount fucking Everest."

"Almost done." Steve turned to us and winked. When it looked like the man was going to have one of those incidents right there and then on the treadmill, Steve fiddled with the intravenous tubes.

"Okay, you can slow down now," Steve said. "We've released the isotopes, which will make their way to your heart muscles. Then we'll get you into the imager, snap a few family photos, and you can go home."

We followed to the imaging room, which contained a bed with some rounded plates surrounding it. Steve quickly explained, "It looks like a CAT scanner or MRI machine, but in this case, the patient is radiating, we just take pictures of what radiates out of his heart, a lot simpler." A technician was sitting at another Sun workstation. It looked like she was taking imaging files, adding color and then assembling them in rows for presentation to Dr. Bernstein's reading room, which probably had 20 people in it by now.

"Let's go back to the reading room," Steve said. I groaned.

"Okay, next is Mrs. Birnbaum," Dr. Bernstein declared. "Steve, you're up. And no vintages."

"Okay. Lateral a bit weak. Aortal wall thin. Can I see the at-rest images?"

"Sure, pull them up." A few clicks later.

"Yup, thin wall. Something else bothers me. That discoloration in the lateral."

"Okay, enough, this one is flagged as positive." Dr. Bernstein said. "Let's move on, we've got lots more to go." The files were stacked high. And our time was up.

Chapter 6

∎

Do Doctors Scale?

For the life of me, I could never figure out what drives people to want to be doctors. Certainly there is the prestige and cachet. Smart, driven, helping fellow man, etc. In the town I grew up in, the doctors lived in their big homes on the top of the hill, which their new Mercedes were more than capable of delivering them to, while the rest of us lived in the foothills, looking up longingly at their lofty status and accumulating wealth. But that was then.

Today they get the status, but no longer the wealth. That goes to option-gobbling CEOs, actors, money managers, and subterranean-dwelling class action lawyers, or so it seems.

What happened? Who wouldn't pay a king's ransom for a doctor's healing powers?

It's those damn insurance companies, right? The mess inflicted on health care is well documented. Wage controls innocently put in place during World War II forced employers to come up with something to attract workers. Employer-provided health coverage became the norm. As untaxed compensation, it had the indirect approval of the U.S. government.

Then that whole LBJ Medicare thing. If you were eligible for Social Security, we'd pay your retirement and, what the heck, we'd pay your medical bills, too. Great Society indeed.

Back then it was considered compassionate to pay for the elder-lies' health care needs. Now it's gotten out of hand.

Demographics, smoking and the proliferation of Mc-Donaldies over the next 60 years were not kind to health care costs. If you hit 65 in 1940, your life expectancy was 77. If you hit 65 in 2005, you were expected to live almost to 83. A little girl born in 1995, if she hits 65, will probably live to 89. Cancer and heart disease and stroke and obesity are all treatable to some extent or another, as opposed to curable, but certainly not for cheap.

The health insurers that popped up to handle corporate medical spending put their foot down in the early 80s. Managed care became the corporate buzzword. Insurers said, "You can have whatever health care you want, as long it's on our approved list." Health maintenance organizations gave Americans a taste of socialized medicine. Faceless administrators, random doctors, long waits, rationing of limited resources—it felt like the Moscow plan.

By the mid 90s, workers revolted and demanded choice. They got it at the expense of higher deductibles, copays and a nightmare of other bureaucratic hassles. Medicare squeezed even harder. Their reimbursements followed strict schedules, with very few exceptions. Hospital stays got shorter. And doctors got caught in the squeeze, making less per patient, and they all had to go back for additional training—in filling out complex forms. Productivity became the buzzword. If you cut back office visits to 12 minutes from 15, you could see 20% more patients and make the same amount of money. To paraphrase John Belushi, "Twelve years of college down the drain."

As a kid, I got pretty good grades. Family friends who were doctors encouraged me to go into medicine, but I was too busy trying to grow up to give it much thought. At some point in my early teen years, though, I had an experience that made the decision for me. One hot summer day, I was swimming at my friend Charlie's house, when his dad, Dr. Schwirk, ran past the pool. He stopped long enough to say, "I've got an emergency C-section, you guys want to watch?"

Charlie and I looked at each other, quickly nodded yes, and dried and dressed in record time to catch up with his dad walking down the hill to the hospital. Once there, we threw on scrubs and face masks and quietly snuck into the operating room. Lying on the table, with her eyes wide open and legs spread apart, was a sight that would scar my adolescent memory for life (I should say right here and now that Dr. Schwirk was a veterinarian). The bright lights flickered off a scalpel as he sliced opened the belly of the Boston terrier and pulled out a whitish sac the size of a large bag of Lays potato chips. For some reason, I had to squint to make out the next incision. I was able to count two puppies pulled out, like clowns out of a small circus car, before flashes of light circled my head and I passed out on the floor. Charlie dragged me out of the room and pulled off my mask as I gulped for cool air.

I learned there were seven in the litter and that I could cross off medicine as a career choice. I now think Boston terriers are beautiful animals, greet them personally and thank them for sparing me a doctor's life.

But doctors are a proud bunch—it's not about the money, well, sort of. It is about the patient. Doctors will do everything they can for their patients to make them better. It's that Hippocratic oath they take:

> I swear by Apollo Physician and Asclepius and Hygieia and Panaceia and all the gods and goddesses . . .

Oh, wait, that's the original one. Here's the short version of the modern one:

I swear to fulfill, to the best of my ability and judgment, this covenant: I will respect the hard-won scientific gains of those physicians in whose steps I walk, and gladly share such knowledge as is mine with those who are to follow . . . I will apply, for the benefit of the sick, all measures which are required, avoiding those twin traps of overtreatment and therapeutic nihilism . . . I will prevent disease whenever I can, for prevention is preferable to cure . . . If I do not violate this oath, may I enjoy life and art, respected while I live and remembered with affection thereafter . . .

"Dere's de a-auu-da, da valve, an de roight ventrica." A technician named Dina was giving us a tour.

Six of us visiting students surrounded the Acuson machine. I immediately recognized the V-shaped gray image on the screen from the four sonograms I'd watched being performed on my wife for each of our sons. But instead of looking at a cute face and looking away to avoid knowing the sex of the fetus (boy, oh boy), I was staring into Mr. Kruger's heart, beating away.

Dina was taking measurements, humming a pop song of unknown origin. Mr. Kruger had waited patiently in the waiting room (is that why they are called patients?) and agreed to let us watch. It was fascinating for the first few minutes and then was just a blur of moving gray goo. We were in one of probably six small offices containing an echocardiogram machine (some were made by GE), a bed and a curtain. The walls were covered with phone number lists and comics, as in every office I've ever been in. Except these were all doctor comics, with punch lines like "Well, yes, as a matter of fact, I am God." Must be from *The New Yorker*, funny to no one. A few books were scattered on the shelves, but one with a familiar yellow cover caught my eye as a bit strange for a leading hospital—*Heart Disease for Dummies*.

"Dere's his pacemayka, dat white ting in the a-auu-da."
Pretty cool. Sure enough, there was a pacemaker. It looked like
a cattle prod stuck in his heart, which I suppose it was. I was
going to reach for the book but wandered out of the room instead and ran into Dr. Ferguson.

"Can I ask you a question? How much are these machines?" I couldn't help asking.

Dr. Ferguson gave me a strange look, a what-kind-of-
student-are-you? "Around 250 large, maybe 300."

"Wow, how many of these procedures do you guys do a
day?"

"Let's see." He scanned a printout with yellow highlighter
marks on it. "It's a pretty slow day today, it picks up toward the
end of the week. Oh, I'd guess we'll do around 50 echos today,
but that's just outpatient. They'll do at least that at inpatient. In
fact, you should go check that out. There are a couple of students there now, I'll drop two of you off."

"Good morning, sir." I looked over. Oh, no, it was me and
Troy again.

We followed Dr. Ferguson through a maze of corridors
until we ended up on a hallway littered with body-filled gurneys. He led us into a darkened room with a curtain pulled, and
then left.

"What should we do, sir?" Troy asked.

I didn't think it would hurt to peek inside the curtain.

It was quite a sight. Two guys in scrubs, one with a tattoo
that spelled out Travis in a gothic font, were hitting on two very
blond visiting students in white coats. Troy took a step backward as I flung the curtain open.

The subject changed quickly. "So that's how we determine
infarction inflammagamma . . . itation, thank you for asking,"
one of the guys in scrubs started with a smile as the phone rang.
Travis got the phone as a woman in a purple sweater named
Vivian walked in and started bossing everyone around.

"Who's cawlin'?" she asked.

"Dr. Suckme," Travis blurted out.

"Vishna?" purple sweater, whose name tag read Vivian, wanted to know.

"Yup."

"Asshole. Just hang up. Okay, let's get goin', they're backing up in the hawl again."

In the darkened room, a really, really old man on the bed was snoring and tubes came out of both arms while another set of tubes led to an oxygen tank at the foot of the bed. Travis came in with a yellow bed covering stretched across his scrubs, popped in a VHS tape, squirted some gel on the sensors and started the echocardiogram.

"I'm not seein' nothin', Dr. Pete," Travis sighed. It was just fuzz on the screen.

"Gotta move him on his left side," said Dr. Pete.

"I tried, you do it," Travis complained.

"Okay. Shit, this guy probably hasn't moved in three months . . . he ain't budging." The snoring got louder. "What the . . . ? " A stream of something wet squirted all over Dr. Pete's scrubs. Troy and the two blondes backed up faster than I did. "Hey, Travis, is that why you're wearing that yellow thing?"

"Yup."

"Get me one of them, will ya?" Dr. Pete asked.

"Yeah, sure." Travis quietly snickered.

One of the young blond women leaned over to me and said, "When I'm a doctor, no way I'm going to work with old people." I was about to suggest she become a veterinarian when the phone rang.

Vivian got it. "Hey, Travis, it's Dr. Suckme again. He wants to know who you working on."

"Tell him Pedasso. Stu Pedasso." Travis and Dr. Pete started laughing.

Vivian said into the phone, "Travis wanted me to tell you Stu Pedasso." She couldn't keep a straight face. "That's right, Stu Pedasso." I finally started laughing, too.

Meanwhile, the echocardiogram was continuing. I kept noticing some colors on the screen, lots of reds and blues and shades of green.

"What's the colors mean?" the someday-doctor-to-the-young-only asked.

Dr. Pete expanded his chest. "Well, these machines can measure blood flow. You can see it being pumped through the heart. What we look for is regurgitation, some blood flow heading back through the valve. It can be a sign of a bad valve or other disease."

"Do you make that determination?" I asked. It was clear that Dr. Pete was really just supervising and not paying that much attention.

"Somewhat. We output this entire procedure onto a videotape. I'll sit with a cardiologist in the reading room later and go through these one at a time and make determinations with them."

"The actual tapes?" I figured it would all be digitized and put on optical drives like in the cath lab so doctors could quickly get to the good parts.

"Yeah, sure, the tapes. They're talking about putting this stuff on DVD but ran out of money. It works fine how it is, I guess."

"Are you all doctors?" the other young blond woman asked.

Dr. Pete's voice lowered and he whispered, "Well, I am, third-year fellow, actually. Travis and Vivian are technicians, I just make sure it's all done right."

"What do you have to do to become a technician?" she asked.

"Yeah, there's a technical school. You need a high school

diploma to get in. It's a pretty good job, these guys make good money."

The phone rang. I could hear Vivian answer, "The patient? Her name is Aphelia . . ." as we headed out into the hall. Even more gurneys were lined up, as if every patient in the damn hospital was getting an echo today.

I still didn't get it. Who pays for all this stuff? I surfed around between meetings and figured out it's both nobody and everybody. If you are over 65, we all pay for your Medicare via taxes. Medicaid adds to our tax bill the expenses of others unfortunate enough not to be able to afford health care. By 2004, there were 40 million beneficiaries and 900 million claims costing $200 billion.

How do you get productivity out of that? People are sick and dying. Doctors are needed. To me, cutting visits by three minutes doesn't count as real productivity. That's not scale. In fact, I wasn't sure doctors could scale. It's a service business. Each one of them is on their own. If there was some way to package doctors' collective knowledge and sell it in some kit—now, that would count. I just didn't see it. Without Silicon Valley–like scale, I suspected there was no hope. Maybe medicine was doomed to ever-increasing costs and I was wasting my time. It wouldn't be the first dead end I'd marched down.

Meanwhile, it was time for part two of Dr. Blake's talk. He postponed it until after rounds when only ten kids were left, and three of them were asleep.

"Here are some other wacky stats. One third of Medicare beneficiaries have four or more chronic conditions," Dr. Blake started. "5% of the sickest Medicare beneficiaries accounted for 47% of Medicare expenditures in 2002." One would assume that it's in the last month of their lives, I thought.

"But the cliff is coming. From 2006 through 2011, Medicare reimbursements are legislated to drop 5% per year.

That's just fact, we're not at all sure how the industry is going to deal with it, but compensation ain't going up." I notice some fidgeting in the room.

"So, if you are wondering what field of medicine to go into, this chart might help. For family docs, internists and pediatricians, from 1999 to 2003, the total charges, gross dollars that their practice collected, was up by 20% or so. Sounds good, but if you then look at their compensation, dollars in their pocket at the end of the year, it rose barely 8% to around 150 grand. Not that that's anything to sneeze at. The reason for this is that reimbursement rates were cut. It takes seeing more patients just to go sideways compwise."

I seemed to recall that 25-year-old second-year investment bankers made that much and didn't have to wash their hands. Didn't seem right.

"On the other hand," Dr. Blake continued, "noninvasive cardiologists, those who run echoes and nuclear imaging, saw almost the opposite. Over those same years, their charges only rose 9% but their comp jumped 25%, to 350 grand. Start adding in CT scans and MRIs and the numbers go up even more."

Dr. Blake paused. "Why is that? Anyone, anyone?" And then he started shouting, "Because they have the toys—imagers, scanners, whatever. Anyone with toys breaks out of that pay-per-15-minute-consultation quagmire."

He shook his head.

"Only certain types of surgeons make more on average than cardiologists. And even that is suspect, their pay is dropping. It's all about the toys. The other specialists with toys are radiologists, of course, and the gastroenterologists with their GI endoscopies and, you know, colonoscopies and the like. They're a pain in the ass"—ba-dum-bum—"but it's a lucrative profession."

• • •

"What's the point of all these tests, the echoes and nuclear imaging and everything else?" I asked Dr. Simplot, another cardiologist who spoke to us.

"We want to have a look inside, rule things out."

"But in the cath lab, you can see everything, right?" I wondered.

"Well, sure, but there can be complications—not often, but they happen. Punctures, you know. You don't want to poke a wire around someone's heart if you don't have to. Plus, it's not cheap." I mentally canceled my request for the cath lab.

"But do echoes or nuclears tell you much?"

"Yeah, well, sometimes. We often get accused of self-referral, but that's not quite fair."

"What does that mean?" He'd piqued my interest. Self-referral? Finally, some dirt.

"Look, these machines are expensive. Hospitals own them but lots of independent physicians' groups lease them and then get paid when they refer their own patients to get the tests. But you have to realize, these are doctors. They're not in it for the money. They all take the Hippocratic oath, the patient comes first. These tests are important to uncover disease as well as to rule it out."

"But—" I stupidly tried to interrupt his rant.

"And radiologists, don't even get me started. They call it overutilization instead of self-referral, but jeez Louise, it's the same damn thing. They buy these big scanners and charge you coming and going."

"But aren't CT scans or MRIs better at looking inside patients?" I asked.

"Probably." He sighed.

"So why aren't more done?"

"It's the damn codes," answered Dr. Simplot. "Look, doctors are going to do procedures that they can get reimbursed for. There are codes for echoes and nukes. CTs are reimbursed for head trauma, but those cool new multislice CTs are better, but don't have that many good reimbursement codes for cardio."

Multislice? So the technology existed to look inside me more cheaply and safely than the old catheter through the groin, but Medicare and insurance companies hadn't figured out how to pay for it? Technology was ahead of the curve?

"So, is that a problem?" I naively asked.

"Maybe. But, son"—I love being called son—"did you know that heart disease is no longer the number-one killer of people under 85? There were 1.3 million nonfatal AMIs last year. Cancer now has the top spot. 476,009 cancer deaths and only 450,637 from heart disease."

Doctors are pretty good at memorization.

I was going to ask about stroke, but I think he made his point. Cardiology may be flawed, inefficient, error prone, and riddled with self-dealing, but this stuff actually works. These guys are proving that they save lives. That's priceless. But $210 billion? For all this ancient stuff, from the . . . well, all the way from the 80s!

"Actually, deaths from heart disease have been declining since the 50s. No one really knows why. Fewer people smoke, diet is better, more exercise, new drugs—we just don't know. Of course, I'd like to think it's cardiologists."

Yeah, right.

"You've heard about Medicare and the reimbursements cliff and the $1.8 trillion spending number?" Dr. Simplot asked.

"Sure," I answered.

"Well, there's only one real way to solve the problem of runaway spending on health care in this country."

"Okay." He had my attention.

"Just keep people from getting sick in the first place."

Chapter 7

■

Zap It Out

Wilhelm Conrad Roentgen gets credit for discovering X rays. It was another classic case of accidental progress. On November 8, 1895, Wilhelm, a German physics professor, was playing around in his lab with Crookes tubes. These were basically an early version of vacuum tubes that would eventually go into radios and radar and TVs. Roentgen was blasting electrical current through them, hoping to measure how far cathode rays could travel outside the tubes.

One tube was wrapped in cardboard, and when he cranked up the current, a screen across the room that happened to be coated with barium platinocyanide started glowing. That was odd, as cathode rays of electrons weren't supposed to do that. Roentgen started holding up materials between the tube and the screen to see what might block the rays and noticed the bones of his hand in a faint outline of flesh showing up on the screen. In phraseology not as profound as "What hath God wrought" or even "Mr. Watson, come here," Roentgen told a friend: "I have discovered something interesting, but I do not know whether or not my observations are correct." He announced his discovery to the world on New Year's Day 1896, refusing to patent it and thankfully insisting on the name X rays instead of Roentgen rays.

Within two weeks, newspaper in the U.S. announced: "New Light Sees Through Flesh to Bones!" The prognosticators came out of the woodwork—one guy even claimed, "Soon every house will have a cathode-ray machine." He was right if he meant TVs. He may yet be right about X rays.

A few weeks after these headlines, doctors began using X rays looking for broken bones and swallowed coins. If only new medical technologies could be adopted this quickly today.

By the turn of the century, radiology was a branch of medicine. Stanford University hired its first radiology instructor in 1904. 100 years later, they threw a party at a building called the Lucas Imaging Center to celebrate a century of radiology.

Everyone who was anyone in radiology turned out. Jeffrey Immelt, the CEO of General Electric, a big maker of X ray scanners in computed tomography imagers, was there. So was Erich Reinhardt of Siemens Medical.

Dr. Gary Glazer, the head of Radiology at Stanford, talked about multidetector row scanners and where medical imaging was headed. Dr. Sam Gambhir talked about personalized imaging. Talks were given on Hodgkin's disease, breast carcinoma, and on and on. X rays have come along way since Roentgen's bony hand.

Up to the podium strode Dr. Amato Giaccia—a doctor via Ph.D. in a room filled with illustrious M.D.s. A professor of radiation oncology, he gave a talk titled "The Future of Targeted Therapies for Radiotherapy." No one expected it to be a highlight of the celebration.

But Amato Giaccia let loose a quote that turned heads in the room. He claimed, almost in passing, "There is not a single cancer that is truly resistant to radiation. If we can find it early enough, we can eradicate it."

Chapter 8

■

Blood Tests

It was time to see if those Egg Beaters were doing their job. Was my cholesterol dropping? Good up, bad down? I sure as hell wasn't going to drop another couple of hundred bucks with Dr. Greedy. No way—I was off the grid and planned on doing this myself.

So I shopped for a cholesterol test kit at Walgreen's. No dice. All you get is one number for overall cholesterol. Nothing about LDLs or HDLs or MDLs (more damn lies). Surely something exists out there away from doctors. I did some more digging to figure out the ins and outs of blood tests. My conclusion? They are probably the biggest scam going in medicine.

Quest Diagnostics is the big elephant in the business. Lab-Corp isn't far behind. More than likely, your blood tests are done by Quest, which does about $5 billion in tests a year, or LabCorp, at about $3 billion. Their tests for cholesterol or B_{12} levels or platelet count are pretty straightforward chemical tests that cost on the order of pennies per individual test. They put a clinical analyzer in a hospital or large physicians' group and then make their money on reagent rentals, the chemicals used in the tests. Pretty smart—give the razor, sell the blades.

Prices for tests are all over the place and are often not based in reality: $25 for that cholesterol test, which isn't bad, I suppose, but $30 for a platelet count, $55 for a "qualitative pregnancy" test and on and on. Each check mark on the form is another $25 to $50 for a nickel, max, of chemicals and a cut for the labs, the hospital and the doctor. I'll bet the accounting is more expensive than the chemicals.

Fortunately, some of these tests are so simple, you can package them in a snazzy box and sell them at drugstores for under $5. I've seen the blue plus sign come up on pregnancy tests a few too many times. But will all of these tests eventually be over the counter? It would be nice.

There are some tests that are slightly more complicated. Immunodiagnostics use antibodies that stick to other cells as a way of determining what is in blood or urine as a marker for potential disease. Abbott is the big player here, with antigen/antibody reaction tests like the famous PSA, prostate specific antigen, test for prostate cancer. The PSA screen is no more than 80% effective—meaning 20% of men with prostate cancer have PSA scores that are false negatives. Not as cheap as chemicals, these tests probably cost in the $1 to $10 range per test, to Abbott anyway. To you and me, the sky is the limit. PSA tests are $100 for the screen and another $200 for a more detailed followup. There is another immunodiagnostic test for ovarian cancer, CA-125, that is also no more than 80% accurate. The price to you is—as you might guess, a number not quite pulled out of the air—$125 per test.

And then there are molecular diagnostics—DNA tests. You take a cell, break it open, pull out the nucleic acid in the form of DNA and RNA, and then amplify and multiply them to make detection easier. This polymerase chain reaction, or PCR, is used to find missing genes or polymorphism where one of the letters in the DNA code is wrong. This is the test to find Down syndrome, and probably a couple of dozen other prenatal abnormalities. Again, not as simple as chemicals, these are multiple tens of dollars per test, although the patent on PCR is just about expiring, so you can bet there is some super PCR in the labs as the existing tests get cheap.

Of course, none of this helps me. I want to just prick my finger and know within a few minutes whether I should worry about dropping dead from a heart attack any time soon.

Chapter 9

■

CAVE

As I scurried across campus, cutting past the Hotel School, I figured I had just enough time for one more meeting. I was supposed to meet J.T. and Pink at Ruloff's for bloodies at three o'clock, but I had this feeling that they would start without me. I was back for my reunion and trying to balance old fraternity brothers with catching up on the latest and greatest. The six- or eight-story-tall brick building was a lot different from the squat engineering buildings I remembered—building code yellow for Electrical, building code red for Mechanical and building code green for dopey civil engineers.

The building was easy to pick out, but the sign made me stop in my tracks and start laughing—CORNELL THEORY CENTER, FRANK H. T. RHODES HALL. Rhodes was the university president when I was there and I had met him just once. Our fraternity house had some big important anniversary, 100 years on campus or some such thing. Muckety-mucks from "national headquarters" flew out for a lavish dinner and celebration, as lavish as our ratty old house could muster. Our cook Kathy promised to bathe that week.

My friend Shelby got stuck planning the event and did almost nothing, except on a lark he had invited the president, none other than Frank H. T. Rhodes, to attend. That was good

for a laugh until he accepted. Even to this day, these stuffy af-
fairs are good for only one thing. Not known as a group to let a
good deed go unpunished, we leapt into action, dragged out old
leisure suits and bad ties and prepared for battle.

I came downstairs late and was issued a name tag by Shelby.
"Kess, good news, you're sitting at the head table." Great.

Standing around the table were J.T. and Pink (you have
probably guessed his last name is Floyd) making small talk with
President Rhodes, trying not to crack a smile. Shelby and I
strolled over. It was time to inflict pain and turn their straight
faces into stifled laughter and possibly tears.

"So, Richard, what are you studying?" President Rhodes
asked J.T. I thought I heard a snort, then Pink turned to face
away and gritted his teeth hard. Something weird was going on.
As far as I could remember, no one at the table was named
Richard.

I spied J.T.'s name tag and it read "Richard Bagg." Oh,
shit. Pink's said "Richard Hedd," Shelby's "Richard Brane"
and much to my surprise, mine read "Richard Waad." I caught
Shelby's eye and somehow held back a grin, but I was laughing
so hard on the inside, tears were forming. Our sophomoric
humor was starting to backfire.

To give him credit, Frank H. T. Rhodes didn't bat an eye.
After sitting through a few boring speeches, he got up and
wound a story about walking up to our house, which was quite
run-down, and saying that nothing is quite like it seems on the
surface and that labels can often be wrong. He proceeded to
look each of us in the eye and say that while it's not easy, you
have to look on the inside to really understand someone, and
on and on. I think he got the last laugh.

I was still chuckling when I stepped off the elevator on the sixth
floor of Rhodes Hall. I followed signs for the CAVE—Computer
Assisted Virtual Environment. The demo sounded good in the

activities description, but it was past three and my interest was fading. The door was closed, and as I opened it slightly, the room was mostly dark and 30 or so people were sitting around wearing funny-looking oversize glasses. I spied an empty seat in the front row, tripped over a couple from the Class of '55 and sat down, almost crushing a pair of those glasses.

"You want to wear those. You won't get the 3-D effect from that end." The voice came from a guy with a bizarre-looking headset and a short Harry Potter–like wand attached to a cord. He was standing on a 10-by-10-foot floor, and in front of two more 10-by-10-foot walls—three faces of a cube.

"I'll remind you, these glasses are liquid crystal shutters, synched up to the display. We calculate and display images for your left eye and block out your right, and then every 60th of a second flip over, display images for your right eye and block out the left. No more of the red and green glasses from the 50s."

The couple next to me nodded to each other and I thought I heard him whisper, "Like at the drive-in, darling."

"As you can see, I can pull up menus in the CAVE, move them around, and click on what I want with this wand. But what's really cool is visualizing stuff. Here, let me pull up my favorite. Okay, if you have ever been to Florence, you probably waited on line to see Michelangelo's *David*. During its recent restoration, cleaning really, they let in some Stanford researchers to do a 3-D capture of David over a 30-day period, head to toe, tip to stern." Mrs. Class of '55 started giggling. "A set of lasers scanned the entire sculpture, digitizing every nook and cranny and crack." More giggles. "We cut a friendly deal and were able to obtain the very large data set—the Digital David.

"We've got a whole bunch of Windows PCs just above our heads—Microsoft is one of the benefactors of our work. These machines crank away, calculating how every one of the pixels should be displayed for your left and then right eye. Here, let me show you . . . Cue the drumroll."

In a flash, spread across both walls and the floor was David's head—curly hair, Adam's apple and all. Better yet, it was being rotated around, zoomed in on. I've seen lots of 3-D demos, studied ray tracing and Z-planes, invested in 3-D chip companies, and I was completely floored by the beauty of David sparkling on the walls of a 10-foot cube.

"This is a PG show, so we are sticking with just the neck-up view. You're looking at a piece of a couple of billion pixels, probably now about million-plus polygons, each accurate down to two millimeters, which is why I am able to zoom in so closely—let's look up his nostrils—there you are. Now let's back off a bit. Here is something you probably didn't notice if you saw David in Florence. He's a bit cockeyed. His eyes are looking in two different directions."

I'll be damned. Sure enough, one was looking right at me, and the other at Mrs. Class of '55.

"Of course, we can only capture the surface of the sculpture. It's solid marble. The way light diffuses and reflects off the marble is also lost, so we get this dull gray look, but it serves its purpose. Now, you'll notice when I turn him upside down that the model that represents David is hollow, assuming his skin is—oh, I don't know—a centimeter thick. Saves time on calculations. But it also lets us do something fun. It takes some gymnastics up here, but I can usually pull this off, bear with me."

The guy in the funny helmet spun his wand in a circle, which caused the head to rotate around, so we were looking at the back of David's head. Then he squatted down, looked up, and raised the wand above his head. The effect was to look up into David's empty skull from the hole in his neck. You could see the waves in his hair from the inside. Quite cool. But then, the wand was lowered and we were inside David's head, which basically felt as if your nose was David's nose and your eyes were David's eyes.

"Okay, now if I click to lock in this perspective—ah, there

we are, done. Now I am David, fetch me a slingshot, I feel strong enough to slay giants today. Let's look around."

He turned his head left and right, up and down, and David's inside-out view changed—you felt as if you were King David sizing up your opponent. Then the screen went blank.

"Okay, enough of this fun stuff, let me show you what this technology is really used for." Click, click, click.

On the walls flashed what looked like a set of Tinkertoys my tenth-grade chemistry teacher used to play with. Colored balls and lines or double lines between them.

"This is a model of glucose—just to show you we can represent macromolecular structures in 3-D. This is the killer app for pharmaceutical and biotech companies. Here, let me bring up a model of a tumor cell—a bit more complex."

The screens got really messy, colored balls and lines flying all over the place.

"We can look at this a lot of different ways. We can do 3-D images constructed using the 3-D-iso-surface module. We process them in Metamorph by low-pass filtering, and then do multi-thresholding to segment the image volume and distance transform applied to create concentric shells of the cell."

Huh?

"You are probably all saying 'Huh?' Here is the punch line. A scientist can study these models and then invent proteins that fit perfectly into the shape and folds of the cancer cell. This is called rational drug design and promises to bring all sorts of miracle cures. In fact, there is a company in Massachusetts . . ."

I heard some fidgeting next to me and Mr. Class of '55 was mumbling under his breath, "Jeez, this presentation has just entered Fantasyland. Rational drug design is about as real as the tooth fairy. Doris, let's go. It's time for bloodies at Ruloff's."

Chapter 10

■

LASIK

Please take a seat. The doctor will see you shortly." I doubted it. There must have been 100 people sitting in chairs in a makeshift waiting room. As I looked around, I realized that it was just one large room, with a piece of equipment up front that buzzed or maybe flashed every couple of minutes.

My wife was scheduled for her LASIK surgery at 10, but fat chance of that. This place was like Grand Central Station. She had taken the prescribed Valium or some sort of sedative at breakfast, and now had a funny grin on her face like this was the greatest place to be in the world.

We agreed on the procedure without a long discussion. Nancy figured she spent $10 to 20 a week on contact lenses, so the $900 an eye was pretty easy to justify. But more important, it was the 15 minutes of hassle and, how should I put this, mood degeneration, that took place just about every morning getting them in—hydrating, futzing, cursing—that would be hard to put a price on. I turned to talk about this, but Nancy was staring off into space, and the first glistening of a spot of drool was showing up in the corner of her mouth and I knew she was off in never-never land. Maybe now was the time to discuss that 12-cylinder vehicle I had been researching—she looked like she might agree to anything.

"Nancy, great to see you again."

I looked up to see a very odd sight—a 12-year-old boy dressed in green hospital garb, pretending to be a doctor.

"Hiiiiiiiiii" was all she could say.

"I see you have taken your little pink pill," he said.

I gave him my best skank eye.

He turned to me and put on the serious doctor face and said, "We can't have patients squirming around and moving their eyes, can we?"

"Are you the doctor?" I asked incredulously.

"Oh, sorry, I get that a lot. Hi, Dr. Edward Manche. Gotta run, we are backing them up again."

I watched him walk double time to the front, drop a few drops into someone's eye and then place his face into the contraption, whisper something in his ear, and then push a button. Like everyone else in the room, I jumped when the machine flashed. Maybe it was just the lights flickering, like Dr. Frankenstein throwing the switch. A nurse (who was probably 18 but looked older next to Dr. Munchies) led the patient away. I thought I heard a "Next" from the front. Henry Ford would have been proud.

I had done a little homework. The black box was a LASIK eye surgery machine. LASIK stands for laser in situ keratomileusis, which is Greek for "blast high energy coherent light, but only into the cornea." Or something like that—I'm not sure I would trust an ancient Greek with a laser.

You basically numb the eye, open a flap in the cornea and move it aside, then blast an excimer laser to remove just enough corneal tissue, to an accuracy of 1/4000th of an inch, to reshape the cornea and change its refraction to correct near- or farsightedness and astigmatism, which I've heard has something to do with the stigma of wearing glasses.

I leaned over and told Nancy, "Doogie Howser has got a hell of a money machine going here."

"Whaaa?"

"You got a 12-year-old man-child who got out of med school this morning taking in 900 bills every three or four minutes. It's a frickin' racket."

"He niiiice doctahhhh." I wiped her chin.

After a few minutes, the nurse came and led Nancy slowly to the front. I looked around and noticed for the first time a small group sitting separately in the front. There were three or four people with masks over both eyes, like they were on the red-eye to London. Doogie was over chatting to one of them and then ran over to help ease Nancy's face into the LASIK machine. I couldn't watch, but a minute later, she was being escorted back to me with a patch over her left eye.

"Flawless," exclaimed Doogie, who then lowered his voice and said to Nancy, "Call me if you have any vision problems, tunnel vision, blockiness, out of focus, it's rare, but let me know," and then he raised his voice again, "and I'll see you in a 10 days for your other eye. Good luck."

"Ka-ching," Nancy said quite coherently. The lease on the half-a-million-dollar machine and a couple of hundred dollars in royalties for each laser blast means Doogie clears more than every cardiologist and brain surgeon in this hospital. Not bad work if you can get it. Damn those Boston terriers. If that's all you have to do, maybe I should have been a doctor.

But it's more than that, isn't it? If Doogie is blasting a hundred eyes a day, that's got to be millions of dollars a year out of the pockets of optometrists and ophthalmologists and folks at Pearle Vision Centers, let alone drying up sales of Bausch & Lomb saline solution. A new piece of technology in the hands of a specialized doctor could mean literally dozens of other doctors out of a job—the bank tellers of a new era.

Chapter 11

■

The Big Three

When I mention I'm interested in technology and health care, I usually get a lecture on electronic health records, EHR. Fix the back office of medicine—put medical records on-line—and you've solved all of its problems. I'm not so sure. I've seen enterprise software work for corporations, but mostly to help them grow more profitably, not to cut costs and save the company.

I've watched lots of companies get strangled implementing too much technology too quickly.

Then there's GM. Nothing EDS and Ross Perot did could save the beast at General Motors. It had bigger problems that a digitized back office couldn't solve.

Lots of folks want to attack this problem. Steve Case of AOL fame is trying to bite into this problem—something called Revolution Health Group. He's got Carly Fiorina, ex of Hewlett-Packard, Franklin Raines, ex of Fannie Mae and even Steve Wiggins, ex of Oxford Health Plans. Their plan is to buy things and put companies together and try new approaches. Perhaps mailing out zillions of CDs or funny accounting might be tried next.

I just didn't see the point of digging into the IT of medicine and finding out that it sucks. So what? It will get fixed over time. United Airlines moved from sorting paper tickets to airport kiosks and electronic tickets. They still filed for Chapter 11.

I just didn't see how electronic medical records were going to change the fact that we spend $1.8 trillion on health care and are quickly headed toward $3 trillion. The patient is dying. No time for cosmetics.

One of the reasons I just didn't see IT as the solution is a few numbers that got stuck in my head:

- 70% of health care spending is on chronic disease.
- 5% of the sickest Medicare beneficiaries account for 47% of Medicare expenditures.
- $210 billion was spent last year on cardio and stroke, $192 billion on cancer.

Heart disease and cancer are battling it out to be the number-one killer in America. Death rates from heart disease have been dropping, from almost 600 per 100,000 Americans in 1950 to under 200 today. Cancer, on the other hand, has been dead flat over those same 55 years, 193 deaths per 100,000.

In 2003, the National Cancer Institute announced their Challenge Goal—to eliminate the suffering and death due to cancer by 2015. The administration and just about every member of Congress have "affirmed commitment to ending cancer death and suffering." Voters love it. Of course, in 1971, Nixon declared a war on cancer and signed the National Cancer Act, and not much has changed. Somehow, I don't get the impression you can eliminate disease by legislative fiat.

So heart and cancer each kill close to 200 per 100,000, and stroke comes in at 50 deaths. Those are the big killers. But there are plenty of other diseases. There's obesity, diabetes and arthritis. There's Alzheimer's and that thing that Lou Gehrig died from—plus hundreds of rare diseases that affect children and adults. There are lots and lots of chronic diseases. The only one that is important is the one that you have or your family and friends have. In other words, there are a lot of moving parts in medicine.

I had to narrow down my search or I'd drive myself crazy; $1.8 trillion was just too big a number to dissect into little pieces. If I was going to find something, some technology that changes health care in any meaningful way, I needed to stick with heart, stroke and cancer. I had to follow the money. I had to stick to the Big Three.

Chapter 12

■

Magic Pill

And then I started thinking. Bear with me here—it doesn't happen all that often.

What if there was some magic pill, a cheap wonder drug, discovered tomorrow that would both reduce plaque in arteries and kill all cancers? You'd take it every day, preferably with a glass of single-malt Scotch, and poof, no more heart attacks, no more stroke, no more cancer. The Big Three go bye-bye.

What would medicine look like then? We'd still need emergency rooms for gunshots and kids who stick quarters up their noses. We'd need ob-gyns and birthing rooms, and maybe a couple of dentists, but the whole damn health care industry would be turned on its head, wouldn't it?

We'd need fewer doctors. Actually, a lot fewer doctors. The drug industry would have to do something else for a living. It would be a pretty different world.

Besides more nursing homes for all the old folks who haven't moved to Las Vegas, without heart attacks, stroke or cancer, that $1.8 trillion spent on health care in the U.S. would probably halve. And then halve again.

We'd see an uptick in vacation homes, a glut of veterinarians, new anxiety drugs, cigarette and cigar sales booming, and lots of life insurance salesmen handing out towels at all the new golf courses.

Of course there is no magic pill. I'm just thinking.

Chapter 13

■

Meeting Gary Glazer

Years ago, back when I was a serf/analyst following tech stocks at Morgan Stanley, I would occasionally come up for air from the depths of techdom and Silicon Valley and try to figure out how the rest of the world works. The firm had some great analysts—I wish I'd had more time to listen. Michael Sorrell knew more than anyone about biotech. Paul Brooke was the expert on Big Pharma. I listened but didn't really get it.

One slow day, I plopped my butt down in Steve Roach's office. He was Morgan's economist and I figured I could bother him to see if I could learn anything about the mumbo jumbo he dealt with—GDP, inflation deflators, floating currencies, consumer sentiment, whatever.

"Steve, I snort bits and bytes and megahertz all day. If I stare at them long enough, I think I can see that every time they go down in price, new markets are created."

"You mean elasticity, right?" Steve asked.

"Yes, but—"

"Could be. It works sometimes, you may have the only sector. Everything else is inelastic. Cigarettes, oil, autos, health care, you name it. They go up in price every year and they sell more. Investors, rightly or wrongly, flock to this stuff."

"And that's the economy? What exactly is an economy for?" I asked. I'm an engineer, I slept through Econ 101.

"Real simple," Steve answered. "The economy is nothing more than a mechanism to increase the standard of living of its participants. Period. End of story. The rest is noise. You happen to be following a business that contributes to productivity and can help the economy grow and wealth be created. Other things like health care, well . . . that's what we spend our new-found wealth on. It all works."

And there you have it, a postgraduate degree from Steve Roach in less than five minutes. He threw me out of his office as the weekly payroll numbers hit the tape.

I focused on the words *increase* and *standard*. Technology does that in spades—it's productivity in a box. But over time, I kept thinking back to that definition and the word *living*. You've got to be alive for this whole thing to work. That's what health care is. Keeping your ass alive and maybe even well. As Steve told me, "Technology creates wealth and health care is what you spend it on." True, but is that too simple? Maybe technology can be part of "living," not just "living standard."

"Hey, Andy, I notice Steve Roach is speaking tomorrow at a Stanford economic forum. You want to go? You can sit in the back and kibbitz." It was Fred Kittler, my old partner from Velocity.

"Do I have to wear a tie?" I asked. I have very strict rules.

"I doubt it. You should go."

Steve Roach wowed the crowd at Stanford. There were a bunch of venture capitalists and even one or two Nobel Prize winners in attendance. The conference room at the Lucas Center was overflowing. I was the only one without a tie. My being there was a surprise for Steve, who recounted a few old lines from some pieces he and I had coauthored way back when, about technology and the economy, blah blah blah. Mostly, he warned about the coming global rebalancing and labor arbitrage and I started nodding out.

"What are you doing here?" Steve came up and asked at the end of his talk.

"Slumming," I answered. "Figured I would step in front of a tomato or two aimed in your direction."

"Funny. I see you haven't changed. Speaking of tomatoes, there's a dinner at John Bentley's for a few folks—you should come. I think Fred's going."

I looked over and Fred was nodding his head and saying, "You should come."

And this was how I found myself in a private room at John Bentley's, sitting next to the head of Stanford's Institute for Economic something or other, John Shoven.

After some chitchat and wine pouring, John pointed to the nice gentleman across the table and asked me, "You know Don Lucas?"

"Sure, yeah, everyone knows Don," I said.

Don Lucas was one of the godfathers of venture capital. He had funded National Semiconductor, one of the companies I followed, but made serious scratch in Oracle and a few other big names like Cadence and Macromedia. He'd made a fortune a few times over. Of course, it hit me right then that the Roach talk was at the Lucas Center. Aha.

"You know Steve Roach from . . . ?" Don Lucas asked me.

"Morgan Stanley. Many years ago," I answered.

"Yeah, yeah. I remember you. Semis, right? Followed National?"

"That's me," I said.

"And now you're making an honest living?" Don asked.

"Sort of. I'm pretending to be a writer," I admitted.

"Oops," Don Lucas said with a smile.

"I'm trying to see if health care can do the same thing semis do—you know, get cheaper, create new markets."

"Well, you need to talk to Gary over here," Don said, pointing to the gentleman sitting next to him, who was engrossed in a conversation with someone else. With that, Don got up and started chatting with the other table.

I leaned over to our host, John Shoven. "Who is that guy sitting next to Don Lucas?"

"Oh, he runs the other Lucas Center at Stanford."

"There's more than one Lucas Center?" I asked.

"Don had a brother, Richard, who died much too young. Don felt that technology might have saved his life, so he endowed the Richard Lucas Imaging Center. This was probably in 1992. The gentleman next to Don is Dr. Gary Glazer, who heads up Radiology at Stanford and runs the center."

"Sounds like the guy I've been looking for," I said to no one in particular.

The night was dragging on. Steve Roach had finished answering questions. Dessert was served, so I felt it was my right to interrupt Dr. Glazer and pump him for information.

"Dr. Glazer—" I started.

"Gary. Please call me Gary."

"Gary, I've been searching and searching health care for anything that scales. Smaller, cheaper, faster, better."

"We can do that," Gary told me.

"You can?" I was somewhat taken aback. "What do you mean?"

"That's what we're working on at the Lucas Center. All sorts of imaging systems. Every year they get faster and better. The machines are more expensive, for now anyway, but the scans get cheaper and cheaper. It's changing radiology. You're looking in the right place."

"I am?" I was still surprised.

"If you're really interested in this stuff, you can come by, but in a week or so, we're holding an event up in San Francisco—Multidetector Row CT," Gary said.

"And that's?" I interrupted.

"That's the 20th, I think," Gary answered. That's not what I was asking, but I didn't want to sound too stupid. I usually wait until a second meeting to unload my ignorance.

"I'll put your name on the attendee list," Gary continued. "Give me a card. We can talk afterward. I think you'll be impressed. The old days of black-and-white X rays on a light board will soon be behind us."

"Uh, thanks," I kind of spit out. The dinner was over but I think I just blew my schedule for the next year or so.

Part II

Chapter 14

■

The Dish

When I think of health, I think of two interrelated things. The first is living forever—100 probably works just fine for me. The second is not twisting in pain gasping for life during a heart attack or chemo treatments. I'll meet my maker when I'm good and ready, on my own terms and with no false alarms along the way.

But how? Maybe I'm lucky and won the longevity gene lottery. But maybe not. If not, would I be happy to trade some of my wealth for health? Whether any of us likes to admit it or not, we set the alarm for 5:45 A.M. and commute 50 minutes to get yelled at by some asshole boss, just so we can bribe the grim reaper (via our doctors) and don't die early from some painful disease.

But it's odd. You can be disgustingly rich and go to your doctor and say, "I'd like to live to 100" and all you'll get back is lay off the Krispy Kremes and consider Egg Beaters. Good luck with that.

As Steve Roach taught me, economies are about increasing standards of living. After the La-Z-Boy, plasma TV and cable are paid for, and, well, throw in beer, wine and few warm vacations if there's any money left, we'll all trade a few shekels for health and well-being. If only there was something to buy.

$1.8 trillion is spent on health care in the U.S., but very little of it is bought. But it has value. It is a business. And as a business, it's subject to market forces, innovation, change, disruption and obsolescence. It's just no one wants to admit it.

Meanwhile, with high blood pressure and high cholesterol, according to the literature, anyway, I needed to get my heart pumping, to flush all that crap out of me, I supposed.

From what I know, those looking for a good outdoor cardio workout on the mid-Peninsula south of San Francisco usually do one of two things. The first is get on a bike and do The Loop—a fierce 28-mile climb and descent on Sand Hill Road, Alpine Road and Skyline Boulevard, finishing with a wild downhill dash on curvy La Honda–Woodside Road. If your tires don't blow out, sending you over a cliff, there is a better than even chance you'll be run over by a Ferrari 360 Modena taking the same turns faster than you. I thought I'd pass.

A safer cardio workout, much more my speed, is The Dish—a four-mile paved loop through a fenced-in set of grassy hills, right past every radio-astronomer's wet dream, a 150-foot-diameter, 150-ton parabolic reflector capable of one-degree-per-second tracking sitting on top of a hill with beautiful vistas of the San Francisco Bay. Nancy and I hike this all the time. Far more serious types run it. Ouch. The dish itself is quite impressive.

What they don't tell you is that this dish was originally built to detect radio emissions from atmospheric nuclear explosions. Just as the final nut was tightened, the Atmospheric Test Ban Treaty was signed. Oops. It quickly found other important tasks, such as measuring intergalactic hydrogen clouds at L-band, communicating with and commanding the low earth-orbiting spacecraft system in UHF, and who knows how many top secret projects.

As for me, I always appreciated the wide open space—I'm talking hundreds of acres here—on the western side of Stanford campus. On one of our hikes, I overheard two über-geeks behind me talking about the Bracewell Observatory and solar flares and what a shame it was that they were going to tear it down and something needed to be done about it, it was a crying shame, blah blah blah. I will say that it sounded like a "Save the Whales" sort of thing and my mind is programmed to tune these things out.

But as the week progressed, I started seeing stories in the local newspaper about the imminent destruction of the Bracewell Observatory, and just how important Ronald N. Bracewell was to humankind. So against my better judgment, I started paying attention.

It turned out that down the hill from the big mother of an antenna on top of the hill, there are a number of small antenna farms that were first installed in the early 60s. Go down the hill, across some rickety bridge and along a dirt road and you come to five 60-foot dishes, not quite as famous as the ones on the hill. Stanford's fire inspector was concerned the place would soon spontaneously combust and wanted it torn down. I'd heard of this happening—it's so dry in the summer that everything has a dead brown look to it. He condemned the area, known as Site 515, and a movement to save it soon jumped into action.

Just what is so interesting about some rusty dishes in Site 515? The bottom of one article mentioned Bracewell's involvement in CT scanners.

Now, that got my interest, so I dug around and found out that not only was Bracewell still alive and well but a professor emeritus with a phone number and email.

"Just what's so interesting about some rusty dishes at Site 515?" I asked Professor Bracewell.

"I prefer the term elderly over rusty," he told me.

"Okay, fair enough."

"It's all in the history. In the late 50s, I joined as a scientist at the Stanford Radio Propagation Lab. It eventually turned into Stanford's Space, Telecommunications and Radioscience Laboratory. STAR Lab for short.

"We were running all sorts of experiments on radio waves, terrestrial and outer space, but I focused on the sun. In 1956—remember, this was a year before *Sputnik* prompted the U.S. to enter the space race—I was building dishes to map the sun's surface.

"Not big ones like up the hill. I used 32 smaller dishes, nine meters in diameter, and arranged them in a cross pattern, 7.5 meters apart. All together, it was one arc minute, if you know what that is."

I nodded and grunted, remembering a high school astronomy class, but all I could come up with in my brain was something about how many barleycorns per foot-arc minute were in Avogadro's number. Best keep my mouth shut. I grunted again.

"We had the first antenna to reach the angular acuity of the human eye."

I definitely had to grunt at that!

"A single motor and gearbox and two 375-foot-long driveshafts moved each dish in unison so we could track the sun across the sky.

"I finished it in 1961. You can't believe how hard it was to place dishes that precisely, surveying with geodetic precision, compensation for velocity of wave propagation in waveguides.

"These dishes began taking measurements of microwaves emitted from the surface of the sun. I had the data, but my goal was a daily microwave spectroheliogram, basically a weather map of the sun's surface. This map could see active regions that might produce solar flares before they arrived on the visible hemisphere of the sun, flares that might disrupt shortwave radio communications. The air force was all over this.

"Remember," Bracewell emphasized, "we didn't have

computers back then, just slide rules." I thought he was going to call me a young whippersnapper.

The story continued of life pre-computers. He probably could have painstakingly plotted out all the data by hand on graph paper and then calculated at what angle the waves left the sun so as to hit each discrete dish, then used that angle to correct the data for each dish, and then summed up all the calculations for each potential wave on the sun's surface and generated one map by the year 2015. Five 30-foot dishes were added, at various arc minutes apart, and even more data was streaming in.

So like any good scientist on a budget and an Aussie at that, Bracewell sat down and started working out a solution on the cheap, an algorithm that would cut his computations to a minimum. This is what Fourier transforms are for. You have to go back to your college calculus class to remember that a Fourier transform is an integral function that converts frequency data into a time or spatial representation.

The basic math that Bracewell started with was first worked out by Austrian mathematician Johann Radon in 1917—a Radon transformation that integrated data along parallel lines to reconstruct an object.

Bracewell needed just the surface of the sun and introduced the idea of a filtered back projection algorithm. Think of seeing just the shadow of a tree over time and then working backward to what the tree looks like in real life. If you had the pattern of waves after they went through the cross section of an object, you do an inverse Fourier transform, and back into a pretty close approximation of what the object looks like, without doing all that much math.

Bingo.

The newly formed NASA found out about it and had to have it. President Kennedy had them putting a man on the moon by the end of the decade, a full employment act for physicists, but a big concern for manned space exploration was radiation from solar flares. These microwave bursts would be as dangerous for an astronaut as putting a sheep in a microwave oven—Johnny Carson's Carnac the Magnificent's question to a card that read Sis Boom Bah.

With Bracewell's Observatory cranking out maps quickly using Bracewell's shortcut, NASA could warn astronauts of solar flares almost in real time and have them hide behind radiation shields on spaceships, abort space walks or quickly head inside the Lunar Module instead of hitting golf balls on the moon.

"The sun maps, which were delivered daily by teletypewriter to NASA and the air force, were produced by a pencil beam raster scan, like television."

The Bracewell dishes stayed in operation for 11 years, the length of a complete solar cycle. All these dishes were then more or less abandoned in 1972.

But Bracewell's algorithm was just warming up.

In 1972, an engineer in England, Godfrey Hounsfield, and a South African physicist working at Tufts University in Massachusetts, Allan Cormack, each independently used Bracewell's filtered back projection algorithm. Or so it would seem.

Thanks to our old friend Roentgen, X rays have been used in radiology since 1895. You blast X rays at a body and film is exposed to the X rays that make it through. Bones are opaque, so no X rays get through them, while organs absorb some of the X rays, so they show up in various shades of gray.

What Hounsfield and Cormack each did was take a series of X ray images while the X ray tube was rotated around the body. Then, using Bracewell's algorithm—working backward from the shadows over time, if you will—they could create an image that was a close approximation of a slice of the patient's body. It's as if you were a magician on the *Tonight Show* and took a hacksaw and cut the body into slices yourself, without the box. A sleight of hand, but this was remarkable stuff.

Tomos = slice; grapho = draw. Computer axial tomography, computer-assisted tomography, computed tomography, they're basically all the same thing, CAT (or CT) scanners.

"So they used your filter-back stuff?"

"Hounsfield claimed that he did the reconstruction by solving 30,000 simultaneous algebraic equations, but I suspect this was to impede the competition that would probably arise from GE, Varian, Xerox. He might just conceivably have used my work," Bracewell told me.

"Might have? Sounds like he must have," I said. "It would be hard even today to solve 30,000 equations in real time," I noted.

"It's not really known if Hounsfield used my algorithm. The 1967 *Astrophysical Journal* is the one containing the back-projection algorithm. It is arrived at using Fourier reasoning, but the end result requires only arithmetic, which is why it has turned out to be so useful."

"And Hounsfield split the Nobel Prize for it," I said. Hounsfield and Cormack split the 1979 Nobel Prize for Medicine.

"Sure. Cormack didn't work with Hounsfield. He was at Cambridge at the time, as I was. His claim to half the Nobel Prize was to have worked as an X ray health technician in South Africa monitoring the exposure of operators. The Nobel committee didn't think that a mere engineer like Hounsfield, even though he was a genius, deserved the Nobel Prize in Medicine all on his own. Even though he did."

"Shouldn't you have gotten a piece of it?" I regretted asking the question as soon as it came out of my mouth.

"A lot of people knew that Cormack's citation described work that I had done earlier. That is why we were both nominated at the same time for membership into the Institute of Medicine of the National Academy of Science. Personally, I never thought the algorithm was what I regarded as Nobel quality, so I never felt bad about it. Alan Cormack was just an ordinary physicist like me."

And so it seemed, a modest one, too.

Without his algorithm, it would have taken forever to reconstruct all of that data into anything useful, but now, faster than a heartbeat, images of human slices can be drawn, turning the field of radiology on its head.

I'm one of those folks who snidely believe only two good things ever came out of space exploration—Velcro and Tang (and the latter is a tad questionable). But to anyone who will listen, I'll bend over backward to tell them computed tomography is the greatest thing NASA didn't quite invent.

"By the way," Bracewell told me as we were winding down, "after computers arrived, I kept my slide rule and six-figure logarithms in a drawer in my desk, in case the power went out. They're still there today."

Little did he know that ditching his slide rule would so radically change medicine.

At some point in the 60s, it was discovered that too much exposure to X rays was linked to leukemia. Hospital workers draped themselves in lead shields and the flashes of X rays were kept to a minimum.

Radiation comes from a lot of sources. The average person is exposed to 360 millirems or mrems every year. Radon in the air is about 200 mrems. Food exposes you to another 40 mrems.

Is this a problem? Those who handle radioactive materials are considered safe by international standards if they are exposed to 5,000 mrems each year. An X ray for a broken arm is no more than 1 mrem, a chest X ray about 6 mrem.

Your TV set spits out 1 mrem per year. Airport security is 0.002 mrems.

When you fly in an airplane at 36,000 feet, you are exposed to 0.5 mrems for each hour of flight. A six-hour flight is like having a few X rays. I've heard the dirty little secret that airline pilots are susceptible to leukemia, perhaps because of repeated exposure.

A heart and body CT scan exposes you to 110 mrems. That's a significant number compared to a chest X ray. It won't put you over the top, but you don't want to do one every day.

A bigger issue with CTs is the fact that soft tissues, like the brain, pancreas, kidney and most of the central nervous system, don't show up well with X rays. They don't absorb enough of the rays to provide contrast on the image. This can be aided with contrast agents, dyes, that can darken organs so they show up on images, but only somewhat.

Another technique first played around with during the same early 70s time frame as computed tomography is magnetic resonance, exciting water molecules in just the right way. No X rays.

The protons of hydrogen are distributed differently in different tissue types. If you can crank up a strong enough magnetic field, and blast pulsed radio waves into tissue, you can get these positively charged hydrogen nuclei to point in a certain direction and record the hydrogen density.

Magnetic resonance imaging is the ultimate noninvasive imaging system—you turn on the electromagnets and an image is produced with brightness as a function of the hydrogen concentration and intensity. This is especially helpful at the edges of tissue, when it changes from one type to another—say, a cancerous tumor. MRI didn't hit its stride for human imaging until the 80s.

Chapter 15

■

Close to Home

Why couldn't my doctor use these scanning tools? I mean, all I really wanted my doc to do was to take a look inside—sniff around a little bit—check my arteries and prostate, look for aneurysms and cancer growths. Was that too much to ask? I mean, they do it on *Star Trek*. Bones just waves a Tricorder over Kirk in the sick bay and has an instant diagnosis. C'mon, we already have much better communicators than Kirk and Spock, why not body scanners?

I'd gotten enough tours of factories, wafer fabrication facilities and corporate cubicle cities to have an eye for where all this was going. I look at a screen of nuclear images and realize it will soon be animated. I look at echocardiograms and wonder how long it will take to digitize and send to server farms running regression analysis to find defects. I sit watching a stent inserted via a two-dimensional screen and foresee the day when two imagers reconstruct the arteries in 3-D, making navigation with a catheter less treacherous.

After that week in the hospital, I was dying to know if there was a better solution to all this. They can look inside us, but unless they push in a catheter or crack open our ribs, the information is fuzzy.

But it's just data. Surely some technology exists to turn all that fuzz into something useful, some visualization, some detailed imagery of our innards—oh, look, a collapsed artery. Surely our doctors deserve something as useful, exploring our arteries, shooting clots, avoiding white blood cells (oh, wait, that's the movie *Fantastic Voyage*). But why not a virtual Fantastic Voyage?

We all know someone who has had a heart attack. My brother-in-law Bob Wysocki, 46, is just a few years older than my wife, in shape, healthy, fun loving, a few cigars now and then. Then one night he has a heart attack, is rushed to the hospital, given aspirin and then stents to support his arteries. He's fine, but now has a lifestyle change, diet, pills, some blood thinner (he can't remember what kind) and whatever else. What caused this? A stressful job? Was it genetic? Who knows? But it all seems so preventable if someone had bothered to look inside and sniff around.

Or my old friend Andy Huffman. He grew up in the Pennsylvania Dutch country—played high school basketball with one of my heroes, Tommy Herr, the second baseman for the St. Louis Cardinals in the 80s. Went to Delaware, and had to cover Ralph Sampson when they played Virginia. We worked at Bell Labs at the same time, and played Ultimate Frisbee. Sounds dorky, but I never ran more or was in better shape in my life.

Here was someone in great shape—a big, strapping stud of a guy. I used to hang around with him at parties and more than one tipsy reveler mistook me for him late at night. Nuff said. Now there's a friend.

I lost touch with Andy and then his name came up as a CEO candidate for one of my old investments. He fixed it and sold it and then took over a company named Audible just in time for Frank Quattrone to take it public in 1999. The guy had it made, a wife and three kids, money in the bank and then— poof—he dropped dead of a massive heart attack while playing basketball one weekend. I read about it on Monday morning. I think of him every time I lace up for my weekend hoops.

Chapter 16

■

Cholesterol Conspiracy?

So my cholesterol was high. Was it a big deal? Was I going to have a heart attack any time soon? Being pain averse, and having just dropped way too much money on a plasma TV, I really, really wanted to know.

Just about every article on cholesterol cites a *Journal of the American Medical Association* article from 1984, Issue 251, pages 351–74. It is the original "literature" that my doctor and every other doctor talks about. 1984? How appropriate.

Fortunately, it's been updated a few times since 1984, via a series of phone-book-thick documents known as ATPs, Adult Treatment Panels. The latest is the Third Report of the National Cholesterol Education Program (NCEP) Expert Panel on Detection, Evaluation, and Treatment of High Blood Cholesterol in Adults. What a mouthful. Isn't there some rule of thumb that the longer the title, the less useful the contents?

ATP III makes a pretty strong case for the link between high cholesterol and chronic heart disease, especially elevated levels of low-density lipoproteins, LDL. But when you read some of the wording, you start to wonder:

The ***robust relationship*** between total cholesterol and CHD found in epidemiological studies ***strongly implies*** that an elevated LDL is a *powerful risk factor* . . . A *causal role* for LDL has been corroborated by controlled clinical trials of LDL lowering.

The emphasis is mine, but I recognize this language. It's no different from that of a Wall Street analyst hedging his or her stock picks—outperform, above consensus, subject to betterment.

It seems that this paper was saying, "Well, we never said if your cholesterol is high, you'll have a heart attack." Oh, no, it's just relationships, implications, roles and risk factors. Puh-lease.

Then again, the data is pretty strong. 12 statin trials of 17,405 patients showed a mean cholesterol lowering of 20% and a 30% reduction in CHD (coronary heart disease). That sounds pretty good, doesn't it?

Trials involving two different statins show reduced LDL and 31% and 37% reduction in major coronary events.

But then I started digging into the numbers. If you're 60 years old, with cholesterol over 240, off the charts in anyone's book, you have a 51% lifetime risk of CHD. If you are only 40, you have a 57% 40-year risk. Okay, I'll buy that, scares the hell out of me. Those are terrible odds. Except, look at it the other way: 49% of 60-year-olds with off-the-charts cholesterol have no problems at all. Instead of taking Lipitor, they should be off to the diner for a dripping Danish and an omelet. Which group are you in? Doesn't it matter?

In 2004, the "experts" updated the ATP III with a few more studies.

An Anglo Scandinavian Cardiac Outcomes Trial—Lipid Lowering Arm (that's ASCOT-LLA to you and me) tested 19,342 people with hypertension and at least three cardiac risk factors. They went on antihypertension regimens, sit-ups perhaps, and then a little over half were given 10 mg of a statin and the rest a placebo. There were 100 primary events, negative outcomes as they say, in the statin group compared to 154 in the placebo group. The reduction was 35%. I suppose what I found so remarkable about this study, which certainly shows the value of this statin, is that 100 folks still had an event of some sort. They can put you on Lipitor or whatever and still not really know if it is going to do away with that damn heart attack. And that whole 35% reduction thing is a tad misleading: Never trust a percentage of a percentage. 1.03% of the statin group had an event; 1.59% of the placebo group. So 98.41% of the placebo group was just fine, at least over the course of the study.

Another new study, ALLHAT (I won't waste space with what it really stands for), tested 10,355 people over 55 with LDL between 120 and 189, something like half with a statin and the rest with just "usual care." Sure enough, total cholesterol in the statin group dropped by 17% versus 8% in the usual care group. But over a very long study period, CHD event rates were considered even. So do statins work or not? I still don't know.

Then came what shouldn't have been a surprise—a financial disclosure of the "expert panel" who put the Adult Treatment Panels together. All outstanding doctors and researchers, to be sure, but check this out as a sample of their interests:

Dr. G. has received honoraria from Merck, Pfizer, Sankyo, Bayer, Merck/Schering-Plough, Kos, Abbott, Bristol-Myers Squibb, and AstraZeneca; he has received research grants from Merck, Abbott, and Glaxo Smith Kline.

Dr. M. has received lecture honoraria from Pfizer, Merck, and Kos; she has served as a consultant for Pfizer, Bayer, and EHC (Merck); she has received unrestricted institutional grants for Continuing Medical Education from Pfizer, Procter & Gamble, Novartis, Wyeth, AstraZeneca, and Bristol-Myers Squibb Medical Imaging; she has received a research grant from Merck; she has stock in Boston Scientific, IVAX, Eli Lilly, Medtronic, Johnson & Johnson, SCIPIE Insurance, ATS Medical, and Biosite.

Dr. H. has received honoraria for consulting and speakers bureau from AstraZeneca, Merck, Merck/Schering-Plough, and Pfizer, and for consulting from Kos; he has received research grants from AstraZeneca, Bristol-Myers Squibb, Kos, Merck, Merck/Schering-Plough, Novartis, and Pfizer.

"Hey, you gotta make a living," is how Wall Street shrugged off their conflicts, but this is life-and-death stuff. $25 billion or more in statin medications are sold each year based on this ATP. Pfizer makes a living off Lipitor. Are they stealing from senior citizens? And are the "experts" collecting money from Big Pharma? This could make even a U.S. senator blush.

So, should I worry about my cholesterol? You're damn right I should. Was I going to have a heart attack any time soon? Don't know. I'd rather not have one, given a choice.

Is there an industry-wide conspiracy to pump statins into Americans on the false premise that if they don't, they'll have a heart attack any day now? Hmmm. These doctors and statin snakes may be killing us. Not literally by giving us something poisonous, but by omission, soaking up dollars best spent elsewhere. You would think the $25 billion we spend on statins could certainly be spent more effectively on something useful. But on what?

Just about every piece of common sense inside of me screamed that there had to be a better way!

Chapter 17

■

Calcium Score

Curiosity killed the cat. I hoped it would have the opposite effect on me. I filled out a form on the Web for a local scanning clinic and my phone rang about 10 minutes later.

"We have appointments available," said some guy named Ram.

"Okay. Uh, can you just tell me a little more about these scans."

"Oh sure, it's pretty straightforward. EBT technology, that's e-Beam, you know what that is?"

"I think."

"Best there is—3-millimeter resolution. We lay you out on the gurney"—it sounded like an episode of *M*A*S*H*—"and hook you up to some probes and then image from rest to rest, so your heart and blood motion doesn't blur anything. We can do it for $445, since you came in through our website."

"And you find?"

"We'll find all the calcified plaque. You get a consultation with Dr. Zander and you'll get a calcium score. We had a guy just last week who we scanned, no symptoms, slightly overweight, blood pressure on the high end, cholesterol borderline high."

I gulped. It sounded like me.

"And we found all sorts of calcium and within the week he was having an angioplasty and two stents put in and now he's sending his whole family and friends in for scans."

"It's noninvasive? No needles or any of that stuff?" I asked.

"No needles. We can do it right through your clothes. And for only $300 more, we can do a complete body scan, lung, liver, kidneys, legs, all that. And for $1,295, we can do a virtual colonoscopy. How old are you?"

"46."

"Yeah, you better get one. Beats a real colonoscopy. You fast for 36 hours and we give you a colon cleansing kit. Pretty straightforward. When you get here, we fill your colon with air to get a better image. We stick a catheter up your . . ."

"Thank you," I said loudly.

". . . and pump in air for a few minutes."

"Yeah, I think I'll stick with a heart scan," I said.

"Great, any symptoms? Previous heart attack or anything?"

"No."

"Too bad."

Man, that's harsh, I thought.

"If you had, we can sometimes can get insurance to pay. More likely if you have a high calcium score on the heart-only scan, you can get your doctor to send you to a cardiologist and then the cardiologist to send you here and then Blue Cross, for example, does a diagnostic review and then usually pays the $2,500 for the full angio scan."

"I hope it doesn't come to that."

"Yeah, of course, me too. Dr. Zander is here Mondays and Tuesdays."

"How about 11 on Monday?"

"Perfect. See you then."

• • •

I didn't sleep well over the weekend. Maybe it was the tequila from the poker party on Friday night, but when I finally did fall asleep, I had this vague memory of a giant heart in front of me, beating once and then stopping until someone threw salt and squeezed lime juice on it. Then I woke up with the room spinning.

Cupertino could as easily be on the outskirts of any major city as in Silicon Valley. Apple Computer is around here, but it's mostly forgettable three- or four-story office buildings surrounded by gas stations and retail strip malls. It was like a tour of virtue and vice. For every fitness center, Jamba Juice and holistic center that I passed—what is a holistic center, anyway?—there was a Taco Bell, Carl's Jr. and cigar store.

I pulled up next to a Mercedes S55 AMG—maybe it was the doctor's—and sat in the parking lot for a few minutes. My heart was pounding, and the butterflies in my stomach were migrating up to my throat. I was having trouble swallowing, let alone breathing. It wasn't too late to bail.

I pulled it together and wandered in. I was the only one in the waiting room and filled out the paperwork to the sounds of some really cheesy radio station, Mix 106 or something equally lame. I paid with my Visa card and then sat down and looked for the children's *Highlights* magazine, but realizing I wasn't at the dentist's, instead found *Sunset* magazine and an issue of *Time* magazine with a cover story titled "How to Stop a Heart Attack Before It Happens." I'll bet they bought out the newsstand.

"Come this way, Mr. Kessler," said an orderly, technician, whatever you call someone in scrubs who doesn't immediately introduce themselves as a doctor.

He led me into a not particularly large room with the

scanner. The machine was maybe 10 by 10 with a round opening and a gurney to lie down on. And a Bounty paper towel where my head was supposed to go. To mop up the bleeding?

The machine said Imatron. They were the scanning company bought by GE back in 2001. I didn't see any General Electric labels anywhere. Hmmm. This thing is ancient.

"Any metal?" he asked.

I took the pen out of my shirt pocket and put it near a sink along with my phone. "Okay, just lift your shirt. I need to put on these three probes for your heartbeat and we're all set. Can you hold your breath for 30 or 50 seconds?" I nodded my head. "Great, this should take about three minutes. Put your hands over your head."

I watched a few lasers light up my torso. He must be aligning the thing.

"Okay, take a deep breath and hold it," Garb-man said.

I held my breath and could feel my heart beating. And then a strange sensation. After each heartbeat, the opening of the machine would move back slightly. Beat, move. Beat, move. Beat, move. I got a little spooked when it stopped moving, until I realized that we were done.

"Okay, you can collect your things and wait out in the lobby."

I picked up my phone and pen and noticed about a dozen ants scattered along the counter. Either the staff had all met for doughnuts that morning or they had their own poker party in the scanning room over the weekend. Not a great sign.

The procedure took all of three minutes but I sat in the lobby for another 30 waiting for Dr. Zander, rocking out to Barry Manilow on Mix 106 and then the damn butterflies started flapping again. Something must be wrong. The S55 was gone, I thought I was the only one here. They must be phoning an ambulance and the EMS team.

I picked up another magazine and flipped through it,

looking for anything to take my mind off the pounding in my chest, most certainly caused by ever-shrinking arteries. I found an article about a convicted killer, always a good diversion. On July 3, 1981, a petty crook, Joseph Paul Jernigan, was in the middle of stealing a microwave oven from the home of a 75-year-old night watchman, Edward Hale. As Jernigan was sneaking out, Hale surprised him and Jernigan quickly stabbed Hale and then blasted him three times with a shotgun. I'm drawn to these kinds of stories and quickly forgot about my potentially sclerotic ticker.

Jernigan was caught, the book was thrown at him and he was sentenced to death by lethal injection. It took almost a dozen years of appeals and clemency requests, but at 12:31 A.M. on August 5, 1993, potassium chloride dripped into his veins.

Jernigan had agreed to will his body to the Texas anatomy board to spare his family the cost of a funeral. Pretty thoughtful. What Jernigan probably didn't know was that within 90 minutes, his body was pumped full of five gallons of formaldehyde and then whisked to a plane bound for Colorado. There it underwent X rays, MRIs and CT scans. (Hey, wait a second, what kind of article is this?) Within a day of his injection, his body was packed in dry ice and stored in a meat locker.

In February 1994, Jernigan was cut into four big blocks, blue latex was used to fill in hollow cavities like his lungs and stomach, and then he was systematically cut into 1,878 cross-section slices, each one mm thick. The slices were quickly digitized and put on a CD-ROM and eventually became part of something called the Visible Human Project and were put on the National Library of Medicine's website. I was just about to turn the page to look at pictures when . . .

"Mr. Kessler?" a voice interrupted.

"Uh, that's me." Not that anyone else was there. At this point, I was happy to put down the article.

"Hi, I'm Dr. Zander," a nice 60-something-year-old man said, extending his hand. I looked for a sign of anything on his face. Nothing. Poker cold. Damn.

"Come on in." He led me to his office, which had a huge LCD monitor.

"Thanks."

Dr. Zander clicked a new window open.

"Here is what's known as a slice. This is you." I checked out a bunch of gray blobs on the screen with some white ovals surrounding it. Thank you, Bracewell, although I couldn't quite figure out what I was looking at. Maybe it was Jernigan!

"Kind of a top-down view. Ribs here, spine, breast plate. That blank space is your lungs. Not much there, just a lot of hot air."

I was in no mood for jokes.

"Now, I can take the scroll wheel on the mouse and move through these slices. Each one is one and a half millimeters. It's not hard to find your coronary arteries. Okay, here is your left anterior descending, and left circumflex. Gray is good. No white spots in any of them. Pretty clean."

I could feel my heart rate slow down and the pounding stop.

"Left main, right coronary, see them? That was the proximal view, and we can do the same for mid and distal."

I stared. I squinted. I couldn't see a damn thing—just gray goo melting from one slice to another, like a psychedelic light show going on behind Jimi Hendrix at the Monterey Pop Festival.

I started looking at other stuff on his screen. The software was from a company named TeraRecon. Never heard of them.

I also noticed a box labeled Agatston Score. Wasn't that the South Beach Diet guy? Just about everyone I knew had bought his book and tried the diet, most lasting well into the second week before giving it up for something else.

"Well, good news, Mr. Kessler. I don't see signs of coronary calcifications. Your Agatston score, you know, calcium score, is low—you're doing fine. I'd say a 5% chance of any coronary heart disease."

"Wow. Whew. That's a relief. Still 5%? I can live with that. But what about my blood pressure and cholesterol? They are borderline high. Shouldn't I worry about that?" I asked.

"Well, high cholesterol is linked to coronary atherosclerosis, so they say. But no one really knows how much. In all my years, I've seen plenty of people with low cholesterol have coronary disease and plenty of people who have high cholesterol and LDL numbers never have any heart problems."

"I don't get it," I said. Maybe he could fill me in on the conspiracy.

"Look, no one had these scanning tools when cholesterol studies were done. All they had back then was rates of heart disease—but only after the fact. People who had heart attacks tend to have high cholesterol, so there is a risk, it's just hard to quantify. It would be nice if they looked inside and know definitely."

It would be nice. People spend $25 billion plus a year on statins and other cholesterol drugs, never mind drugs for high blood pressure, all based on the word *risk,* which may or may not be real.

"What about soft plaque? X rays can only really pick up on calcium, right?" I had done my homework.

"True. There are other tests for that. Newer scanners. Cath labs and an angiogram. We can see only calcium."

"So what do you think I should do about my blood pressure and cholesterol?"

"Look, coronary calcification is one of those progressive things. Even if you had a higher calcium score, my advice would be to monitor it over time. If it doesn't change, then you're probably okay. I'd come back in five years and run one of these scans again, when you're 51 and then when you're 56, and maybe even when you are 61. If nothing shows up then, you can stop bothering. I'd start worrying about dying of something else."

Comforting.

I could cross a heart attack off my list, but that left the Big C and a stroke still lurking.

Still—good-bye, Egg Beaters; hello, rib eye. And cigars. And Scotch.

Economists call this a moral hazard. I never liked economists anyway.

But that's it, isn't it? Maybe the jig is up on the cholesterol conspiracy. Any real scientist running studies on cholesterol drugs would not just check to see if participants in the study had a heart attack. You would scan, check for plaque, provide drugs, scan again, see if the plaque increased or decreased, repeat. Instead, we have a multibillion-dollar statin business based on vagaries and deception.

Chapter 18

■

Not Medically Necessary?

I pushed 1 for English, 3 for reimbursements, 4 for clarifica-tions, was asked for my contract code and social security number and then I got the music. Barry Manilow, yuck. Does everyone pump in Mix 106?

"I'd like to get reimbursed for a medical procedure," I said when a human voice replaced Barry.

"Name?"

I gave it.

"Social security number?"

"I already punched that in," I protested.

"Social security number?"

I gave it.

"ID number? . . . contract code? . . . mother's maiden? . . . effective date? . . . Okay, sir, how can I help you?"

"I'd like to get reimbursed for a medical procedure," I said again.

"Do you have the procedure code?" she asked.

"Not that I know of."

"Do you have a localized authorization code?" she asked.

"What is that?"

"Do you have any codes?"

"My receipt says 71250-TC/HT," I told her.

"Computed tomography to detect coronary artery calcifications?"

"Yes, that's it." Whew . . . I thought I was going to have to key in more numbers.

"Have you had an AMI?" she bluntly asked.

"I once drove a BMW."

"Sir, have you had or are you expected to shortly have a heart attack?"

"Jeez, I hope not."

I thought I could hear her exhale exasperation. "Sir, those tests are not covered."

"Not covered?" I asked incredulously.

"Policy number RAD dot 00001."

"But why not?" I asked.

"Those are investigational. Not medically necessary."

"Not what?" I was shouting.

"Not medically necessary," she repeated.

"Isn't that the point? I got the test before I had a heart attack," I insisted.

"Sir, there is nothing I can do. Company policy."

"But you're my insurance company, wouldn't you rather I not have a heart attack?" This was going in circles.

"Sir, that is our medical policy, dated April 28, 2005."

"So you're not paying?" I asked pointedly.

"We are not authorized to reimburse for these procedures."

"Well, what if my policy was not to pay my bill, because it's not medically necessary."

"Sir, I need to remind you that these calls may be monitored—recorded for your safety."

"My safety? Ha. Well, thanks for nothing."

"Anything else I can help you with, sir?"

"My eye hurts when I drink coffee . . ." I started.

She hung up. If she had said "take the spoon out," I would have stayed a lifelong customer of Blue Cross of California. But it sounded like I needed a new insurance company.

A few clicks later and I had my answer:

POLICY STATEMENT

Investigational/Not Medically Necessary:

The use of electron beam computed tomography (EBCT), helical CT or multislice spiral (also known as multi-row detector) CT (MSCT) is considered **investigational/not medically necessary** for the detection of coronary artery calcium, including, but not limited to the following indications:

- as part of a cardiac risk assessment in asymptomatic patients
- as a diagnostic test in patients considered at intermediate risk for coronary artery disease, where other cardiac tests have been inconclusive
- as a diagnostic test in symptomatic patients

Assholes. Their rationale was lame. A U.S. Preventative Task Force report, which Blue Cross probably paid for, and an American College of Cardiology/American Heart Association Expert Consensus Document—probably made up of doctors who want to prescribe Lipitor—was the "primary basis" of their conclusions. Jeez. How about this for a primary basis: you're cheap bastards! There, I said it.

I took two of my boys to the dentist that morning. No X rays, no cavities, a teeth cleaning that an electric toothbrush could do better, and the receptionist wouldn't let me leave without paying $245 each. Yet Blue Cross wouldn't reimburse me the $445 to make sure I wasn't going to have a heart attack any time soon? What if the future was here with no one to pay for it? Why is this so backward?

Chapter 19

■

Scan Scam

Shit. I got my answer. There it was in black and white. It felt like air was being let out of my tires.

I thought I was onto something. Real scale, I figured that this thing was a layup. CT imaging might constantly need faster processors, more memory, finer resolution, a classic Silicon Valley tale. I was going to do some more digging. Dr. Glazer's Multirow something or other conference might give me some clues. Maybe, just maybe, if I found what I was looking for—Scale with that capital S—we'd see 20 years of growth with this thing. But now, while I was walking in from the end of my driveway at 6:30 in the goddamn morning—boom—it was completely discredited.

There, on the front page of the *New York Times*, the headline blared, RAPID RISE AND FALL FOR BODY-SCANNING CLINICS. The article declared that after some initial excitement about them in 2001, scanning centers were in decline, the whole enterprise under question and suspicion.

AmeriScan was one of these scanning clinics, founded by a 34-year-old Stanford radiologist, Dr. Craig Bittner. That name sounded familiar to me—then I remembered being inundated with his pitches of his wares on radio commercials.

One of his clinics opened up next to the Neiman Marcus

at the Scottsdale, Arizona, Fashion Square. He did 1,000 scans in his first six months. Passersby could watch the scans being done through a plate-glass window. Neiman Marcus had mannequins, AmeriScan had real live humans. Both, it turned out, were selling overpriced fashion.

CT Screening International also went bust, despite doing 25,000 scans in four or five years. And now, like that, a declaration by the Gray Lady that no way these body scans would ever work.

This article was the autopsy for a business knocked off in its youth. The scanning scheme had all sorts of problems that a good clinician would have identified much earlier. For starters, patients paid. That's not all that bad, except that with a few exceptions, insurers refused to pay. At $1,000 a scan, it took these clinics a few years to make their way through those wealthy enough to afford it. Even at $500, when the business got competitive, business slowed to a trickle.

What turned patients off?

Too many false positives—if there was some artifact in the imaging, a lesion here, a nodule there, patients went into a tizzy.

Dr. Barnett Kramer, of the National Institutes of Health, was quoted in the article as saying, "For every 100 healthy people who undergo a scan, somewhere between 30 and 80 of them will be told that there is something that needs a workup—and it will turn out to be nothing."

Those words were like a stake through the heart.

It seems as if doctors were inundated with scared folks checking up on what at the end of the day were either just computer glitches—some interpolation of two data points that shows up as a lesion or tumor but is just a ghost—or a benign and harmless feature we all live with.

For every story of someone who found a blockage or some aneurysm that was about to rip open, dozens spent weeks feeling nervous only to get more sophisticated CT scans or MRIs and found to be just fine, thank you.

Doctors bad-mouthed scanning and digital imaging from these outpatient hooligans, although it probably didn't help that doctors weren't getting anything out of it themselves.

Maybe it was time for me to go back to Wi-Fi and broadband policy. This was depressing. Silicon Valley economics could tackle telecom. Maybe medicine might have to wait for another day. I still had a few folks I was scheduled to meet with, but my heart was no longer in it.

Chapter 20

■

Rad or CAD

Okay, nothing in that one, let's keep going." Dr. Solon Finkelstein was cranking through these at a pretty good clip. About a minute each.

The light board seemed like something out of the 1970s. Films were clipped in two rows to some white flexible plastic. Dr. Finkelstein would hit a button and a motor would whir and the films would move to the left, wound onto some spool buried in the machine, and a new set of films would roll into view.

"I've been doing these since 1967, you know. Not much has changed. Oh, the film is better, but the rest of this . . ." He waved his hands in front of the contraption.

To his right were color-coded files, the same ones I remember seeing in my pediatrician's office in, well, 1967. Dr. Finkelstein wore a head-mounted magnifier, one that you might a see a jeweler wear. He would occasionally lean forward to look closely at the films.

"Some like to hold up a magnifying glass, but I like to keep my hands free for the papers and the dictation machine," he said.

After peering at the films, Dr. Finkelstein would push a button and a view of the films would come up on two small monitors that seemed to be jerry-rigged to the light board. He would then read a bar code on the patient's records—how 80s—and push a button on a phone and say into a microphone; "Patient Smith, negative, 12-month review, Solon Finkelstein." Two marks with a pen on a sheet of paper and then "Okay, let's go on to the next one."

I stifled a yawn. I had somehow neglected to tell my wife I was going to be checking out women's breasts all morning. I think she'd understand. Staring at mammograms with a 70-year-old radiologist is probably not considered, er, titillating.

"Okay, here's something." I shot up in my seat. Finally, some action.

"Looks like some calcification there, no problem." Oh. "But what's this?" He took out a red grease pencil and circled what he was looking at.

"Granular tissue, perhaps?" I was glad he was thinking out loud so I could understand what was going on, but I remembered that most doctors tend to think and talk simultaneously. It can be unnerving.

"Let's see what R2 thinks." He pushed a button and those two tiny monitors lit up. They had the same identical views as the films we were looking at, except there were a few black triangles and asterisks on them as well.

"Uh-huh, R2 sees it, too. These triangles are just calcifications, but that asterisk means it might be a mass. Yup, we agree. Usually do. It might be a mass, let's check last year's film. Might be something. No, this is probably no problem. See, it's the same size, it hasn't grown. No problem. Good."

He picked up the microphone, reached over and pushed the button on the phone as I had seen him do a dozen times already. "Patient Jones, benign." Pause. "Let's schedule a six-month follow-up. Solon Finkelstein. Next." That took about two minutes. He was slipping.

"The Bio-Rad classification book is over there, that big

thick one, if you want to go through the government regulations," he said with a smile.

"No, thank you," I said.

"The government insists that if there is something, but less than 25% what I consider granular tissue, I have to call it 'Probably Benign' and follow up in three to twelve months. But who wants to get a report that says 'Probably Benign' and worry all that time? It's a judgment call, but it's often better to say benign and follow up in six months."

The next set of films rolled into view.

"Oh, goodness." Dr. Finkelstein sat back a bit. I noticed that the image on film was twice the size of the others and seem to run off the page. "Well, the extra ones must be here somewhere, usually they are in the file—oh, here it is. Yeah, sometimes we require an extra shot to facilitate the, er . . ." I got it. Pamela Anderson.

"I feel like going home and getting my catcher's mitt," I said to Dr. Finkelstein. I thought he would like the Woody Allen line. Instead, I got a disapproving glance. Okay, no more jokes.

"Some vascular calcifications, no problem. Looks clean. Okay . . ." The two small monitors lit up. "R2 seems to agree, just calcifications, no masses."

I had invested this Saturday morning at the Palo Alto Medical Foundation to learn what radiologists actually do for a living. I figured I would discover some radical change in digitizing mammograms and viewing them on a LCD monitor from electronic medical records, but something else entirely seemed to be going on here and it confused the heck out of me.

"Doc, I promised I wouldn't get in the way, but you keep referring to what R2 thinks. Is that another radiologist? Radiologist 2?" I asked.

"Sort of. Here, I've had enough of this for now. Let's take

a walk, I'll show you around and maybe you'll get an idea of how this works now."

He took off the magnifying headset and scattered the papers around and I quickly followed him out the door. Doctors both talk to themselves *and* walk with a quick pace.

"Okay, this is the mammography machine. You can figure out for yourself what goes where. We take a top-down and a side view, four films in all. A tech brings the films over here to this Kodak machine where they are developed." I could hear Beavis and Butt-Head whispering in my ear, "Did he say developed, heh, heh."

"It takes all of 90 seconds and the films end up in this bin. The tech just moves them over here to the R2 box and loads them in."

"This is R2?" I asked. "It looks like a copy machine." Films were fed in on the top and ejected to a bin at the bottom. An LCD display on top had a few bars moving across the screen to show progress. But progress of what?

"The R2 scans all our mammogram films. I think it takes 30 seconds per image. I look at the original films and then I can pull up what R2 thinks on those monitors. We used to do double blind—two radiologists reading every mammogram, that's the law. Now, it's just one and R2. They call it computer-aided detection, and we can do away with the extra human reading. To me, it's a pretty good backup," he said with a certain satisfaction.

This was new to me—and definitely something that could scale. A doc in a box.

"We do a lot of mammograms here. This is a slow morning and I'll probably go through 40 of them. They've been underpaying for mammograms for years."

"They?" I asked.

"Oh, Medicare, insurance," he said.

"Got it."

"It gets kind of expensive to have two of us read films, so one plus R2 works. Maybe even better, not sure. Plus we get re-imbursements—$29, I think."

"This is pretty new?" I asked.

"We got one of the first ones. R2 is down in Sunnyvale or Mountain View or something."

Figures, I've probably driven by them a million times, imagined D2 across the street and figured it was a robotics company.

"Most mammograms are negative. I think something like one half of one percent show actual cancer. We find that younger docs"—he looked at me like I represented one of those Young Turks—"recall something like 10% of patients for biopsies or MRI or something else. That's a pretty high false positive rate. Us old guys are at around a 4% recall rate. Hey, I've done 250,000 of these over my career and never been sued."

This is about the last guy I would imagine embracing new technology, but here he was asking what R2 might think. Amazing. Then again, his job looked pretty dull to me. Negative, negative, negative. One out of 25 recalled. One out of 250 is actual cancer. That's a lot of negatives. It's like counting cars on the highway, you are inevitably going to miss a few.

Bells started going off. I thought I'd just had a eureka moment. It doesn't happen very often, but I've learned to recognize the sensation. Here was a piece of technology that not only augments doctors but potentially replaces them.

I couldn't help but ask, "So, do you see a day when R2 would look at films first and then alert you to look only at the ones it flags."

I thought he would roll his eyes and harrumph or something, but cool as a cucumber Dr. Finkelstein answered, "I don't think so. Technicians do that already—flag something they see on film, put a mark on it. But the liability is so great, I don't see your scenario happening. I've never been sued, but you need someone to blame." He paused and pondered. "Except maybe in the military, you can get away with anything in the military."

There was another long pause. I thought I'd just found my first use of technology, real bits and bytes and processors and gigabyte drives that not only get cheaper every year but are changing health care for the better. Not like those discredited scanners.

"This is a very efficient process and mind-set. It's all about the experience curve."

Interesting choice of words.

Silicon Valley is definitely one up on Hollywood. They have six degrees of separation from Kevin Bacon, but around here, it is usually no more than three degrees of separation. Steve Strandberg, an old colleague from Morgan Stanley who had his own boxing matches with Frank Quattrone, was an investor in R2 Technologies. Within the week, I was sitting with John Pavlidis, their CEO.

"This detection stuff is pretty cool. So, are radiologists shaking in their boots?" I asked.

"Not yet, but our system keeps improving every day. At first we met with huge resistance, until a landmark study came out in the 90s that showed physicians missing 22% of cancers. In other words, they would get 78% correct and have 22% false negatives. That's pretty startling."

"But how do you do better? I guess I still don't understand how this works. Is there some secret sauce that runs this thing?" I wondered.

"You know about neural networks?"

"I've read enough about them."

"We use them to do pattern recognition. You can code the algorithms, which isn't easy, but the trick is to train the system so that it minimizes false positives but doesn't miss anything. That's tough to do."

"But how?" I still didn't get it.

"Oh, we have something like 5,000 cases of mammograms with known cancer, you know, ones that have proven to be actual positives with biopsies. We train our system on two thirds of those and use the other third to test it out—a control group. You have to do this right."

"And it's cost effective."

"You probably know there are 40 million mammograms done each year, maybe 200,000 cases of breast cancer. Almost everyone has done double-blind reading. That gets real expensive, these are six-figure radiologists. We deliver a computer that gets cheaper every year. It's almost not fair."

"And it's as good as doctors?" I asked.

"We think so. Of course, we run lots of studies to show it works. I think the latest shows something like 7% to 19% more cancers detected."

"And that's enough?"

Someone walked in the door and said, "In medicine, that's huge."

"This is Ron Castellino, he's been our chief medical officer for the last seven years. Ron headed Stanford Radiology and chaired Sloan-Kettering's department."

"From academia and research to a start-up?" I asked. "That's unusual."

"It was inevitable," Ron shot back. "I researched detection for lots of years, but by 1999, I knew that it was finally ready for prime time. I helped R2 get approval from Medicare for reimbursements, and that's when things really took off."

"We've got something like 1,800 units out there, 8 to 9 million mammograms a year."

"So you guys have 25% of the market?"

The CEO jumped in. "More than that. Our competitors . . . you've heard about iCAD, they're public . . . they and others have another 3 to 4 million mammograms. So maybe a third of mammograms use CAD. Remember, there were zero mammograms read with CAD in 2000. This is a huge ramp-up in just five years."

"It sure is," I said.

"We'll get the rest, Andy, it just takes time to change old habits. But the studies continue to prove it out," John continued.

"Breasts are attractive." I caught myself. "Let me rephrase that. Anything else you can attack with your algorithms?"

Ron jumped in, "Any feature that a radiologist is looking for, we can use CAD to find it. We can find nodules in the lung. Lung cancer spreads fast, so we think our stuff will gain acceptance. No one does routine screens of lungs like they do mammograms, but when imaging gets cheap enough, they will and we'll be ready."

"And those guys doing virtual colonoscopies?" John continued. "They call us just about every day asking for help, because they waste all sorts of time winding through folds and turns when we can easily just get them to a feature they can analyze." I wasn't sure what he was talking about. Virtual? Folds and turns?

"Oh, and one more thing. When this all goes volumetric . . ." He paused. I must have looked puzzled.

"You know, 3-D—when it goes 3-D, we really shine. We can do complex segmentation, separate out the chest wall and veins and leave just the organs you want. It's more data, more preprocessing, but then our neural nets really shine. This really can change things in a big way."

Chapter 21

■

Doctors Out

As I thought about it, that really is amazing. From zero to 12 million mammograms scanned in not even five years. Maybe I was being too pessimistic. Maybe technology really can sneak in and just grow into health care quicker than I thought. This is quite remarkable.

And I'm sure Dr. Finkelstein is mistake. It won't take long for computer-aided detection to prove itself out and become the first read—not the backup for doctors. For this to really play out, to lower costs and get better reads, Dr. Finkelstein would probably be asked to read five mammograms a day, not 50. Films flagged by R2 as potential cancer, at $29 per read.

The days of double read are over—an antiquated process to protect us from senile radiologists. We don't need and probably can't afford doctors as the front line of defense anymore. We'll give you a holler when there's a problem. Maybe we really can start getting doctors out of medicine. It's no magic pill, but . . .

That's the problem, isn't it? Doctors. They take a decade or more to train, and then you still have to feed them. Kind of like the Lipizzaner dancing stallion. A neat trick, but terribly expensive to maintain. Okay, more like postal workers. It's pretty neat that you can train people to sort mail, but email servers do it much quicker and cheaper.

And then it hit me. There's a huge difference between medicine and the real world. Almost every other industry is about research and development creating products that are consumed by paying customers. Tons of money is spent researching products, whether that's a Prius, Word for Windows, a Wi-Fi chip or a Motorola RAZR phone. The cost of that research is divided among millions of customers. Wealth comes from innovation—a never-ending regimen of research and refinement and more research until customers form lines out the door.

Medicine is a service business. One on one. You need a doctor, you pay for her time. You need a blood test, you go through your doctor. You need a procedure—your doctor recommends a specialist. With a few new exceptions like LASIK, there is no product—nothing you can go out and buy. Like it or not, you're the product, not the consumer. Even drugs, the one true intellectual property product in medicine, are controlled via doctors' prescriptions.

Doctors hold the expertise. It's embedded in their brains. Only doctors can be Ghostbusters. They are the Gatekeepers, the Key Masters to almighty Zuul. Back off, man, I'm a scientist.

But in other industries, the expertise is increasingly embedded elsewhere—in software, in silicon, in routers, in cell phones, in iPods, in Xboxes, in search engines. That's what made Silicon Valley what it is today. You can take intellectual property and embed it on a chip—to handle telephone calls, move email around, display 3-D graphics for video games and on and on. A dozen guys with no life design the chip and then workerless factories in Taiwan stamp them out by the millions to be shipped in products you and I can buy for under $100. That's scale.

There aren't 10 million nerdy-looking, khaki-wearing Ph.D.s reading search requests at Google. Instead, search expertise is embedded in Google's algorithms on servers in cool dry places. That's scale.

Yet here is R2. Cancer-identifying expertise is embedded in an algorithm you can buy for $29. Well, you can't buy it. Some weird system of service and reimbursements pays for it. But it's the first crack in the armor. It was the first thing I'd seen in medicine that could scale. Doctors, take note. The Geeks are at the gate.

Chapter 22

■

3-D

I figured I had a lot of work to do before I went to Dr. Glazer's conference. As Bracewell had explained to me, computed tomography scanners take a series of X-ray images of your body. They do it in slices, much as those old anatomy textbooks look at a slice of the human body. Radiologists, like my heart-scanning guy, then pour over these images and look for anomalies, calcium, clogged arteries, cancer, aneurysms, polyps. It's not easy, the fuzzy image problem again.

GE makes CT scanners. So do Philips and Siemens and Toshiba. It's not much more than an X-ray machine on steroids. But unlike echocardiograms or that rubber hammer whacking my knee, CT scanners are being constantly improved.

A single slice used to cut it. But I saw it with my own heart scan—radiologists need more, a couple of slices in a row, to make out real organs. Is that a shadow or a real tumor? Successive slices would make all the difference. If only those damn patients would stop breathing and stop their hearts long enough to rotate that X-ray tube around enough times to capture a few more slices.

No problem. You could either move or aim the X ray at will—the issue was really the detector.

I started reading about these multidetector row scanners. It seems the giant breakthrough came along in 1998—you put four detectors on the other side of the X-ray source and you can construct four CT slices for each rotation of the scanner. Depending on how thin the slices are and how many rotations are needed to scan, say, the heart, you end up with lots and lots of images.

Which is great, more is better, but all of a sudden the light board is filled with dozens and dozens of black-and-white images.

You could do what my guy did, arrange them into a little movie and move back and forth through the body, but the real value from all of this data is the ability to create a 3-D model.

Maybe cardiologists, radiologists and someday even me at home will be able to fly through my body. Like that animation we were forced to watch in fifth grade, *This Is Joe's Body*.

Billions are spent on PC games and nVIDIA chips and Nintendo consoles chock full of cheap 3-D chips. Is this a case of leisure outpacing science?

One of the benefits of overpaying for homeownership in Silicon Valley, besides 75-degree Sunday afternoons in December at 49er games, is that if something is new and exciting, it is probably within a 15-mile radius of downtown Palo Alto. Not everything, of course, but I've never been disappointed in searching adjacent zip codes for cool stuff.

I started searching for 3-D models from CT scans, and sure enough, in 94033, San Mateo, was one of the hot companies in the field, TeraRecon. Wow, I should have figured this out sooner. Sometimes something right in front of your face is invisible. It was the same TeraRecon whose software figured out my calcium score and manipulated my scanning slices.

I sent an email at 6 P.M. to the CEO and the head of marketing and by 10 A.M. the next morning I was sitting in their conference room chatting away. Not the first time this has happened, nor the last.

I explained my week with cardiologists and my recent scan, and wondered out loud whether the *New York Times* front-page article meant I was wasting my time.

"You went to the wrong hospital. The leading-edge ones don't bother with all those stress tests and even echoes. 16-slice and certainly the new 64-slice scanners change the game. That EBT stuff you did is four or five years old."

Great, I've been had.

Steve Sandy was an ex-Intel guy now running TeraRecon's marketing. "You need more data, not less. When we get the data, we can do shaded surface display, maximum intensity projection, volume rendering—there are lots of ways to skin this cat. We can shade them, color them, rotate, zoom, fly-through, whatever you'd like. The good news is 3-D algorithms and computer graphics have been around since the 70s—we can take advantage of all that stuff. I'll show you some examples in a second."

I wish I had been to these guys sooner. But I was still skeptical. A 3-D image from crappy gray slices is still crappy.

"Four-slice scanners were the norm a few years back, sounds like your scan was a single slice. Four, then 16-slice, and now 64-slice. Soon, cone-beam scanners that can vary the depth of imaging will offer the equivalent of 256 slices and probably 1,024 slices. The more data there is, the better our system shines."

He babbled on for a while, but I completely tuned out. I'm sure my jaw dropped.

Those were magic numbers he was spouting: 1, 4, 16, 64, 256, 1,024. I've tracked the chip business long enough that I use those numbers to count sheep in my sleep. A doubling every eighteen months, quadruple every three years. It's the metronome of progress. DRAMs for PCs, flash memory for cell phones, hard drive capacity and flash for iPods, megapixels for cameras. 64 meg, 256 meg, 4 gig drives, 16 gig, etc. *Et tu?* And now scanners?

It's déjà vu all over again.

I'd just had eureka moment number two. I hope I wasn't drooling.

I mean, this was exactly what I'd been looking for, something more in medicine that was on the same innovation curves as computers and communications. I think my hands started to shake. I was sitting at the intersection of the video game Halo and sci-fi's *Fantastic Voyage*.

"As I said, the more slices, the more information. These scanners spit out tons of data, and I do mean tons. A megabit per slice per second is probably a decent rule of thumb. Of course, once you get past four slices, radiologists are over their heads analyzing the pure data."

"And that's what your software is for?" I can be a good straight man!

"Sure. We're a 300-person company. We write software to display and manipulate all that scanner data, and we even make our own chips and boards to accelerate the display of our 3-D models. Doctors don't like waiting."

I got a tour of the future. A catheter up the groin has got nothing on the ability to fly through your arteries, looking for bottlenecks. I started to think that with enough compute power and R2's neural networks, that task will even be automated, flagging potential problems for some specialist, hopefully years ahead of when a heart attack hits. Or note that sometimes heartburn is just heartburn. This is the ultimate in prevention as laid out by Master Hippocrates.

Little did I know how much I undershot my enthusiasm.

Chapter 23

■

CT Anxiety

I always feel a certain anxiety when I walk into the Hyatt Regency at the bottom of California Avenue in San Francisco. The cutsie trolley car outside, the Embarcadero tile pattern on the sidewalk—they are all part of the package. But as I've done every time I've been there, I head straight into the lobby, tilt my head back and scan the Escher-like floors, starting at the top and then down and outward to the bottom until I start feeling dizzy. I thank Mel Brooks for this.

With my head spinning from this *High Anxiety* flashback, I strolled into the conference, half expecting to be given a barium enema by a cross between Nurse Diesel from the Mel Brooks flick and Nurse Ratched from *One Flew Over the Cuckoo's Nest*. I really gotta switch to decaf on days like this.

The 7th International Multi-Detector Row Computed Tomography Symposium sounded innocuous enough. I assumed it would be a bunch of technical papers on the future of scanning, which I would read in the darkened hall until lunchtime and then head off for some hot Hunan and home.

Instead, the place was like a carnival for cardiologists. Talk about feeling like a fish out of water. Outside the hall was an expo of sorts, with big signs flashing Toshiba and Philips. Instead of TVs or microwave ovens, there were PCs with 3-D models of some poor schmuck's diseased coronary arteries being folded, stapled and rotated.

The back wall of Toshiba's booth caught my eye and I just stared at it. Rule number one at any trade-show booth is never look interested or you are doomed to a rapid-fire 10-minute lecture on the ins and outs of the product and forced to give up your card as a qualified lead, to be hounded by phone, fax, email and snail mail for the next year.

"Those are our detectors." Damn, I was snagged.

"They look like the display on my laptop," I noted.

"Well, sure, they are not that much different from a flat panel display."

"Same economics making them?" I asked. Flat panels are notoriously expensive to manufacture, because of their size, unlike chips, of which hundreds can fit on an eight-inch-diameter wafer.

"Oh, no, as we go from 4- to 16- to 64-slice, the detectors can be manufactured discretely and butted up against each other. We don't have yield issues."

"How much is one of these 64-slice scanners?" I asked.

"Are you ready to buy one today or this month?" booth-guy asked me.

"No, no, although I wouldn't mind one in my garage. I'm a tech guy."

"Oh, okay. Well, these are basically one- or two-million-dollar machines."

"Wow." I wasn't sure if that was a lot or a little, but often a well-placed "wow" gets you all sorts of inside scoop.

"I know, pretty cheap. We think we have a variety of advantages over the competition and you will see in the face-off that . . ."

"Why so much? I've been in enough factories, and those flat panels are a couple of hundred bucks each and the motor to rotate can't be more than . . ."

"Well, the X-ray source is not inexpensive."

"What? Hundreds of thousands of dollars?" I trolled.

"Probably not. We do have high selling expenses. When you sell only a hundred of anything, there is lead generation and a sales pipeline and funnel."

He started whispering. "They could be a lot cheaper." He must have been having a tough month.

"Don't let me stop you, by the way," I said, looking around, trying to imply he should hard-sell some of these cardiologists and radiologists who were buzzing around the display.

"Doctors aren't buyers, not for these machines. We sell to a few clinics. The rest is into hospitals—they are the only ones that can afford them for now."

"But you said cheaper—I mean, these can be in the hundreds of thousands of dollars instead of millions." It was a statement dressed up as a question.

"Someday," he whispered again.

That's all I needed to know.

I continued my stroll. GE Healthcare was also hawking a 64-slice scanner. I spent some time in front of the Siemens booth listening to a salesman describe how easy it was to fillet a colon. Half a dozen companies from Vital Images to my old friends from last week, TeraRecon, were showing off the coolest 3-D modeling demos I had ever seen, sending other demos back to, well, back to their CAVEs.

Several times, I heard references to the big face-off that afternoon, like it was the reason everyone was there. "Don't miss the face-off," "This ought to show well at the face-off," "This year is going to be so much better than last year's face-off." Okay, I got it.

I escaped into the main conference hall and was handed a program for the next couple of days. I noticed that this was billed as a CME event, Continuing Medical Education, that allowed doctors to earn CME credits and frequent flier miles so they could keep practicing. Fair enough. I always assumed CME was an industry boondoggle with lectures on perfecting your golf swing. But no, this was 150 different talks, each lasting precisely 10 minutes. Again, doctors don't like to give up more than 10 minutes of their time!

Some of the talks looked interesting: 64-Row Scanners: When Is Fast Too Fast?, Future Directions for Cone-Beam CT, The Evolving Role for MDCT in Stroke, Starting a 3-D Lab in a Community Practice. This was why I was here. I figured I would skip the CT of Bowel Obstruction talk given by the aptly named Dr. Chow, especially since it was right before lunch.

But what I was really squirming in my seat in anticipation of was, you guessed it, the goddamn face-off.

Meanwhile, Building a Reliable Bolus, CT Arthrography, CT Urograms: Leak or Full Flow . . .

The head whips woke me up, as my neck turned into Jell-O and my chin dug into my chest. I wasn't sure if I was awake, my heart was beating fast—I was on the top floor looking over the rail next to Mel Brooks . . . Nope, I was okay, I was awake, although embarrassed as quite a few radiologists turned to see what the commotion was in my seat.

My notes were increasingly illegible, but I must have learned a few things. I remember something about 256-slice scanners coming, and something about cone-beam CT, the ability to focus an X ray down to a very tiny beam and adjust its depth in the scan. It reminded me of sitting through electronics conferences, no matter how cool today's stuff was, what a miracle in size and performance were being delivered, there was a bunch of trailblazers looking for the next mountain to attack.

The lights had gone up. There was a break before the big event. I could almost make out a faint chant in the audience: face-off, face-off.

It was time to get some coffee and wake my bleary ass back up.

I strolled through the trade-show area again, being careful not to stare at anything too long and get molested by a quota-carrying booth-guy.

As I took my first sip, I felt a hand on my back. "So, I hope you are finding this useful."

Jeez, I'll buy the damn scanner if you just leave me alone. I turned to look and it was Dr. Gary Glazer smiling at me.

"Oh, hi. Yes, thanks. This is great. Lots of great stuff. I hope some of it sinks in via osmosis." Who was I kidding? I had snoozed through much of it and was now dropping ninth grade biology lingo on a doctor.

"Glad to hear it," he said.

"I enjoyed your talk. I understand your views about how scanning is going to get cheaper and proliferate, but you mentioned something about molecular imaging that I have to admit got by me."

"Well, we are doing some really interesting stuff in the lab, going beyond structural images into functional modality."

"I'm not really following, is it scanning down to the size of molecules?" I asked sheepishly. I hate admitting ignorance.

"Well, we don't want to give it all away at this conference, but once this CT-scanning phenomenon plays out, however many years from now, we need to add biology and in effect use molecular biology to do our scanning for us, provide a signal to zoom in on. The needle in the haystack. This is what personalized medicine will be."

This was all over my head. What could I say except, "Well, I am looking forward to the face-off."

Dr. Glazer gave me a funny grin. "Oh, you'll enjoy it, but you have to look beyond it."

"Beyond?" I asked a little too loudly.

"My view is that heart and stroke disease can be severely

altered with the use of structural scanning. You are right to be at this conference, it's going to be a big business, benefiting from—what's that word you used with me—scale."

"With a capital S. Great." Maybe I was on the right track.

"But that's nothing. We can extend this into the cancer domain and . . ."

He just let it trail off and hang in the air. I hadn't even scratched the surface of this CT and 3-D modeling stuff and these guys were climbing an even bigger mountain. My plan was to follow these folks up the mountain to see what was so exciting.

"I'd like to follow up at some point . . ." but he was gone, probably press-ganged into a diatribe by my Toshiba sales friend.

I moved my seat, having embarrassed myself in the last one, but I also wanted to be closer for the face-off.

"Ladies and gentlemen, welcome back, take your seats, fasten your seat belts, this is going to be exciting. I am pleased to announce that for our 3rd Annual Workstation Face-Off, we have five different vendor groups competing, well, facing off. We have five different data sets, brain, runoff, lung, colon and heart."

The room exploded in applause, like this was some sort of important revelation.

"On the stage, we have workstations from GE Healthcare, Dr. Gruden, please take a bow. Also Vital Images, Philips Medical Systems, Siemens Medical Solutions and TeraRecon. May the best workstation win. Let's get started."

The room was buzzing. Onstage were two giant screens. On the left was a view from the monitor of the workstation and on the right was a live feed from the operator's keyboard and mouse so the audience could see how many clicks and keystrokes and other contortions are needed to get through the data set.

"Okay, let's start with the brain. GE, you have six minutes for both the angiogram and the perfusion. Go."

A giant clock onstage started counting down from six minutes. The doctor operating the GE workstation was furiously clicking and slapping his mouse around, and on screen, we all could share his view zooming through someone's brain.

"Okay, we can see the internal carotid artery on the right-hand side, so now let's quickly move over to this area on the left, ah, not hard to find, there it is, we see the ICA stenosis, let's measure it, 63% blockage." A smattering of applause. "We can zoom in and clearly delineate the calcified versus the soft plaque." More applause.

"Okay, let's quantify the infarct core . . ."

I was transfixed. This guy was zooming through someone's brain like it was a Sunday drive. More like a Sunday afternoon video game. I kept looking for a brain in a jar of formaldehyde labeled "Dysfunctio Cerebri—Abnormal Brain" and Dr. Frankenstein's assistant Fritz limping back to the laboratory.

"Let's mark this tissue at risk for infarction and measure some things while we are over in the left cerebral—okay—MTT is 86.7, TTP, let's call it 52.5 . . ."

He zoomed around the brain like it was level 60 in World of Warcraft.

"Okay. Time: 5 minutes 32 seconds. Very nice. Thank you," the moderator said. The place went crazy. This was repeated on each of the workstations by different doctors to often thunderous applause. I had a mild headache from all the excitement.

I watched these workstations find aneurysms in the arteries from the waist down, the runoff. The trick is to remove the bones from the view and be left with just the arteries. Jeez, everyone knows that. Even I could find the mild aneurysmal dilation of right renal arterial trifurcation! But my feet started to hurt and I looked around and lots of folks were rubbing their calves.

In the lung, the fly-throughs were looking for lobe nodules, which weren't so obvious. It was a maze of tubes in there—who can even find their way, let alone in under four minutes. But sure enough, there was the posterior and the one adjacent to the heart. Each of the five operators then went back and compared them to a study from three years earlier, after finding them in the previous study, of course. Pretty cool. Does my doctor have this? I coughed, more of an unconscious reflex than anything else.

"Okay, a perennial favorite, let's move on to the bowels. This year's virtual colonoscopy will require identifying and measuring five different polyps as well as comparing supine and prone data sets to differentiate stool from polyps."

There was a gasp from the crowd, probably from all the men over 50 who have not so fond memories of their real colonoscopies.

"The folks from TeraRecon will go first."

"Thank you. For this data set, we have decided to show off our handheld interface device. It is a two-handed device, requiring minimal keyboard usage."

On the right-hand screen, the view zoomed to the doctor's hands wrapped around what looked like a Nintendo or Sony PlayStation controller. He was banging it and twisting it around, not much different from my kids playing Halo 2. Except that on the left-hand screen, instead of you as Master Chief blowing away the Covenant to stop them from destroying the earth, you are Master Doctor searching for Cancerous Polyps extracting revenge and trying to destroy your patient.

Or something like that. And you have only six minutes and a crowd of a thousand to cheer you on.

"Okay," the doctor running the TeraRecon station said, "let's go into CAD mode to navigate through the colon."

On screen, he started flying through the wrinkled walls of the colon, twisting and turning, to the left, sliding over, turning up, then right, around a corner, then down again until the scope saw something abnormal and stopped in front of a hanging polyp. Ah, that's what Steve Sandy was telling me about.

Massive applause.

TeraRecon found all the polyps and so did everyone else. It wasn't hard, those polyps hung like fruits from a tree, pretty obvious against the background of the empty colon. Each of the operators had to go to the alternate data set to show that a few potential polyp-looking globes were nothing more than a pile of, well, stool.

My cough had mysteriously turned into a pain in my lower gut.

"Now what you have all been waiting for, the grand finale, someone left their heart in San Francisco."

On screen was a giant rendering of a heart and most of the coronary arteries. It might as well have been pumping and spraying blood all over the audience like in the movie *Carrie*, there was such a frenzy.

Each of the workstations zoomed in, probed for diameters of sinotubular junctions and aneurysmal sinuses. Ho hum. But in no time, each found blockages, stenosis that either had already caused a heart attack or was about to any day.

I just stared at the screen. My eyes were wider than Marty Feldman's as Igor in *Young Frankenstein*. It wasn't some dream of the future, there it was in front of my face. I felt some pains on the left side of my chest, but my stomachache went away.

This was it. The resolution was high enough, and there

was plenty of speed to zoom around and find all the gunk in less than five minutes. These guys could peek inside and tell me if I was going to have a heart attack before I do, before I drop on the floor grabbing my chest and my wife screams to the 911 operator to get someone there as fast as they can, before all my relatives get the call saying Andy has had a heart attack, before I get overloaded with blood thinners and can't remember what day it is.

This changes everything. If I wasn't mistaken, I was having eureka moment number three. Blood pressure readings, cholesterol checks for low-density lipoproteins, echocardiograms, all that stuff is primitive, like silent movies—okay, another Mel Brooks reference. It just has to be cheap enough and it will be as routine as the doctor banging your knee or squeezing the crown jewels.

Let's see, $2 million machine, 10 minutes per patient, of course, that means 72 a day, 360 per week, 18,000 per year, hmmm, that's $111 per scan. Add a little for the attendees and five minutes of the radiologist's time and voilà, maybe this is a mass market thing after all.

Chapter 24

■

Canary in a Race Car

I don't get it," I sighed.

"What's to get? Kevin Kalkhoven owns the whole Champ Car Series. That's what he does now." Lizbeth Moses used to work on the trading desk, and more important, order lunch for me way back when. The fact that she's some high-powered asset manager to the rich and wealthy now doesn't shake that old "I said no mayo" impression I had of her.

"Ah, Kalkhoven. To be an idle billionaire. I remember when we took Uniphase public. What did they do—measure the thickness of disk drives or something like that?"

"It was tough to sell. I think they sold that business pretty quick when they got into optics for WDM for fiber," she recalled.

"But he sold pretty much at the top. Must not have been that big a believer in optical after all," I kidded.

"Anyway, as I was trying to explain, Kalkhoven owns the racing circuit, but this guy Don Listwin owns the San Jose Grand Prix."

"I remember him from Cisco. I suppose he made a bundle. Then he ran that other thing. I remember they merged those two things—Software.com and Unwired Planet. What is it now, something else—OpenWave, right, that's it."

"Yeah."

"Now he races cars, another idler?" I kidded again.

"Not really. He doesn't race. He owns the race, and he's doing it to promote his Canary Fund," Lizbeth corrected me.

"That's what I don't get. What do Tweety Bird and fast cars have to do with each other?"

"It doesn't matter. He's trying to change health care by . . ."

"Now, there's a dreamer," I interrupted.

"No, you should talk to him. You're the one looking for changes in health care. Listwin is working with Stanford and the Hutch up in Seattle, Lee Hartwell, the Nobel Prize winner's cancer work."

"I'm just trying to figure out imaging. I found the coolest company up in San Mateo, TeraRecon. They do 3-D imaging off CT scans. I'm trying to dig up more in this imaging stuff."

"It's all part of what he's doing. Just check it out."

"Why are you so interested?" I wondered.

"You don't know, do you?" she asked.

"Know what? I sold you my co-op on the Upper West Side. I thought you made money on that."

"Yeah, I did. When I was living there, I had some back problems."

"That's not my fault. The floors were crooked when we lived there," I insisted.

"It was a scuba diving thing. I somehow ruptured a disc, so they had me get an MRI, luckily it was nothing. But some radiologist, I don't even know who, saw a shadow on the films. Turned out I had a decent-size tumor on one of my kidneys. I made my way through Sloan-Kettering and they cut me half open to take a quarter of one of my kidneys out."

"Jeez, I'm sorry. I had no idea," I said.

"I'm fine now, but you'd think there would be some way to find this stuff except by accident."

I thought of Brad Miller's skiing and tumor exposing adventure.

"Jeez," I said. I didn't know what else to say.

"Just talk to Listwin. It's pretty different."

Part III

Chapter 25

■

Genetic Tests

I had more work to do with scanning, but if I wasn't mistaken, the head of Radiology at Stanford had more or less told me that heart attacks and stroke might be a thing of the past.

Meanwhile, I was still fuming over my doctor and the $220 blood test. Maybe technology could change the way blood tests are done. After getting ripped off, all I could think about was that old joke about medical tests. I typed the punch line into the search box and this popped up:

A man complained to his friend, "My elbow really hurts. I guess I should see a doctor." His friend offered, "Don't do that!!! There's a computer at the drugstore that can diagnose anything, quicker and cheaper than a doctor. Simply put in a sample of your urine and the computer will diagnose your problem and tell you what you can do about it. It only costs $10."

The man figured he had nothing to lose, so he filled a jar with a urine sample and went to the drugstore. Finding the computer, he poured in the sample and deposited the $10. The computer started making some noises and the various lights started flashing. After a brief pause out popped a small slip of paper on which was printed: "You have tennis elbow.

Soak your arm in warm water. Avoid heavy labor. It will be better in two weeks."

Late that evening while thinking how amazing this new technology was and how it would change medical science forever, he began to wonder if this machine could be fooled. He decided to give it a try. He mixed together some tap water, a stool sample from his dog and urine samples from his wife and daughter. To top it off, he masturbated into the concoction.

He went back to the drugstore, located the machine, poured in the sample and deposited the $10. The machine made the usual noise and printed out the following analysis: "Your tap water is too hard. Get a water softener. Your dog has worms. Give him vitamins. Your daughter's on drugs. Put her in rehab. Your wife's pregnant. It ain't yours—get a lawyer. And if you don't stop jerking off, your tennis elbow will never get better."

With that rattling in my head, I wondered if genetic tests were any good. Maybe there was something in my DNA that would tell me that I was predisposed to some obscure disease and I had better get started on a regimen of Vitamin B_2 supplements and yoga.

Maybe this is the big change in medicine—blame it on your genes. All you have to do is get your exact genetic code and some database exists listing the diseases you are likely to die from.

Plug "Genetic Test Kit" into any search engine and you'll get pages and pages of results. Most of them are for paternity tests—$199 to Genetree and you can get that waitress from Cleveland to quit bothering you at home. For only $245 with free overnight delivery, you can determine if you are Native American and part owner of a casino.

Not having either of those as hot button issues, it took me

a bit longer to surf around and find predictive genetic tests. But I found tons of them.

One company had a test for hereditary nonpolyposis colorectal cancer and endometrial cancer (HNPCC). And another for hereditary melanoma. If I was predisposed to this stuff, I wanted to know. Plus, I could get my parents, whose birthday presents have fallen steadily in value over the years, to pony up and pay for my treatment. With genetic testing, family guilt is finally a two-way street!

I ordered the free test kits, patient packets and DVDs and had them sent to a certain Dr. Weisenheimer, whose offices were conveniently located at my home address.

I found another kit from a different company for cystic fibrosis (maybe I'm a carrier), as well as pancreatitis and colon cancer. Unfortunately, I hit a snag ordering their kit. They started asking some really nosy questions about Dr. Weisenheimer's practice.

Damn. I didn't want to lie to them. It might be mail fraud or some bizarre offense and I figured I might end up in a cell with Bernie Ebbers or Dennis Kozlowski, the worst form of purgatory.

I got stumped on Requestor Type.

Allergist? Nope.
Gastroenterologist? Nope.
Geneticist? Nope.
Infertility Specialist? I've got four boys.
Pulmonologist? What is that?
Paleontologist? Just kidding.
Urologist. Nope.
Sperm Bank. Sort of. It will have to do.

They bought it. The kits were on their way.
About every third or fourth site I clicked on was for DNA

tests for equine and canine parenting and genetic profiling. I learned more than I need to know—there is some immune deficiency in Arabian horses that you can screen for. You can predict coat color in dogs and horses. And lots of dog breeds are predisposed for Type I von Willebrand's disease, and a simple $70 test can let you know whether to overpay for that purebred or if it will surely die of vWD.

What the heck—I ordered one of those kits as well.

Chapter 26

■

Do You Happen to Have a Syringe?

Within three days, I had all three genetic test kits—a kit for colorectal cancer, another for cystic fibrosis, and of course, the Vet Services kit, on the outside chance that I was genetically related to a Boston terrier, to see if I'm prone to some doggie doom.

Unfortunately, there was a slight snag—the two real test kits required that I fill a couple of test tubes of blood. I was going to get a kitchen knife and slice my finger open, but I didn't think I could fill the test tubes before I passed out from the sight of blood dripping out of my finger. I'll get back to these—let's see if my home veterinarian experiment was as difficult.

The terrier test came with four cotton swabs that I was to rub on the inside of my, er, my dog's, cheek. Saliva and all that was in it, hopefully some genetic material, and they could run their test. That was no problem, I had that done in a couple of minutes and the swabs wrapped up and ready to mail back.

The paperwork was a bit tougher. I needed registry, American Kennel Club registration number and exact breed. Registration number of the sire and dam of the tested animal. Then there's DNA profile number, microchip number, tattoo number—what the heck are they doing to dogs these days? Did I miss something? This was going to be messy. But what the hell, I was in deep enough.

Call name: Fluffy
Registered name: Fluffia of Macedonia
Vet: Dr. A. Weisenheimer of the Puppy Love Veterinary Clinic
Registration number: 314159265

Oughta work.

Now I could concentrate on getting all that blood out of me. Can you buy leeches over the counter?

I got in my car and drove to the local drugstore. In the diabetes test section, they had little finger pricks so you can put a few drops of blood into a glucose monitor. That wasn't going to work. A nice old guy in a white lab coat, I assume he was the pharmacist, walked by and I got up the nerve and asked him in the most casual voice I could come up with, "Excuse me, can I ask you a dumb question?"

"Sure, what is it?" He slowed down only slightly as he replied.

"Do you sell syringes here?"

He gave me a scowl, looked at me up and down and said, "You're kidding, right?"

"No, I need it for . . ."

"We haven't sold syringes since the 60s."

"I realize it's unusual, but I have these tests . . ."

"You should probably leave," he said sternly.

"But I just . . ."

"This is a reputable drugstore, we don't traffic in your kind of paraphernalia. We don't even sell rolling papers anymore."

"But—"

He had already walked away.

I walked across the street to Walgreen's but couldn't get up the nerve to ask.

I drove over to my doctor's medical clinic to see if I could get a blood test there. The vampire lab was on the first floor. I walked in and a nice woman asked me if she could help me.

"Why yes, you can. Do you have a list of blood tests that I can get here?"

"Uh, what for? Doesn't your doctor have them?"

"I'm interested in what kind of tests you offer. I'd like to get a blood test. I'm just not sure for what," I answered.

"Talk to your doctor."

"Can't I get a list?"

"It won't do you any good."

"Why not?"

"Because we need your doctor's signature before we can order the test."

"Can't I request them?"

"Sure you can."

"Great. Can I have the list?"

"No."

"But I thought I could request them."

"You can—but only with your doctor's approval. You need your doctor's signature to get a blood test."

I left with my tail between my legs. This whole "take health care into your own hands" thing wasn't working out so well.

Surely there had to be someplace where I could get a simple blood test without paying a doctor to order them, thereby paying a triple or quadruple markup.

I soon found a series of walk-in diagnostic centers scattered around the U.S. I pulled up a list from both Quest and LabCorp, typed in my zip code, and found one less than a mile away on the main drag in Redwood City. So it was back into the car to chase down a simple blood test.

Of course I forgot to write down the exact address. I just looked at a map and remembered the general area and block. And it was Suite A.

I figured I was a block or two away when I saw a sign for Quest Diagnostics—650A El Camino Real. Must be the place. I parked the car and strolled in. The gray carpeting looked like it was put in before World War II. On some folding chairs, in what barely could be described as a lobby, were two guys with slit eyes who looked like Charlie Sheen in his role as the druggie burnout from *Ferris Bueller's Day Off*. The sign at the desk said PULS Drug Screening.

"I'd like to get a blood test," I said timidly.

"Marijuana, cocaine, amphetamines?"

"Sure, I'll take all of them." I laughed. Not a stir. Doesn't anyone have a sense of humor anymore? "Cholesterol, actually."

"This is a drug-testing facility, sir. You probably want LabCorp. They are down the road, 1048 El Camino."

"Suite A."

"Yup, Suite A."

The goose chase is getting wilder.

Two minutes later I was in the lobby of 1048—Suite A. The sign outside said LABCORP—PATIENT SERVICE CENTER and then in smaller letters: ROUTINE BLOOD/URINE AND PEDIATRIC. Finally.

"Can I have a list of blood tests you perform here?" I asked.

"We don't have a list, there's an entire catalog," the woman behind the makeshift counter told me.

"But is there some form?" I asked.

"Sure, but it's blank. Your doctor needs to fill it out."

"Can't I decide and just get a blood test?"

"No way. You need a doctor's signature."

"So I can't get a blood test right now?" I wanted to know.

" 'fraid not," she told me.

"Well"—I was thinking fast—"can I just get blood drawn for some genetic test kits I have?"

"Oh sure, we can do that. Who are the kits from?"

I told her the names of the kit companies.

"Let's see." She scanned a piece of paper with about a dozen names on them. "Oh, here you are, one of them is on the list. We're authorized to do it. Do you have the kit with you?"

"Oh, great. Sure, here's the box." What a relief. Finally.

I unbuttoned my shirtsleeve and started to roll it up as she looked through the papers and the test tube vials that were included in the kit.

"Okay, we can do this, but you need a doctor's signature."

"What?"

"We can do this, but you need a doctor's signature authorizing the test and to receive the results."

"But I want the results."

"You can't get the results. Only your doctor can interpret them for you."

"So I can't just get my blood drawn?"

" 'fraid not."

I bet I could if I were a Boston terrier.

Chapter 27

■

GeneChip—That Counts, Right?

So, do you know of any silicon in diagnostics? I mean, is it really just all chemicals?" I asked my friend Will Kruka, who was in some sort of health care equipment business. I was embarrassed that I didn't know. As with so many of my friends, I knew the company he worked for, knew he ran business development, but had no idea what he really was involved with. More than I thought, it turns out.

We were sitting at the bar at the Tied House, in Mountain View, one of zillions of microbreweries that have sprung up over the last decade. This one was filled with programmers and network administrators and chip designers and search geeks. Still, staring at huge vats behind glass made me think of chemicals, especially C_3H_8O, our good friend Al K. Haul.

"Yeah, it's pretty much chemicals," Will said.

"But what about that GeneChip I keep hearing about. Doesn't Affymetrix use silicon and drive costs down?"

"That's all you digital guys think about, isn't it? Biology is about as analog as you can get."

"But DNA has those four whatever they are called—A, T, G and C."

"Nucleotides. Or bases. Sure. And they pair up to form DNA strands. It's almost digital."

"But it's not?"

"That Affymetrix GeneChip? It analyses DNA sequences and DNA expression—you know, RNA. These chips are probably the closest to a scalable life science technology out there, but be warned, they're very different from a microprocessor."

"How so? A chip is a chip, isn't it?" I asked.

"Well, first of all, it's glass, not semiconductors, and it's got millions of nucleic acid probes bonded to it."

"Glass?" I stared at my beer mug.

"Second, the input to one of these GeneChips is pretty complex. The probes bonded to the chip capture biomolecules containing precisely complementary nucleic acid sequences. Got that? But first, you have to get the right biomolecules on the chip in the right way. The GeneChip basically predefines all the possible results you might be interested in observing." I must have looked confused. "You bait all the hooks with different bait and then see if you caught anything. It's a chemical or biochemical reaction."

Speaking of reaction, our beers were still much too full. I knew Will from when we both lived in Connecticut and he worked for Applera. I vaguely remembered that his company was involved somehow in the Human Genome Project and Craig Venter. I figured if I threw enough beers into him, he'd sing like a bird about it.

"The chip is then optically scanned to figure out which of those million predefined hooks caught a molecule. The chip is a capture device, but then you have to use a sophisticated scanning system to read it. It's just one step in the process," Will went on.

"So it's not a chip?" I was still confused. The India Pale Ale was not providing clarity.

"More of an array. In fact, the industry refers to it as a microarray. The funny thing is, the GeneChip follows its own Moore's law. They can pack more and more features to allow probing for potential events of interest onto increasingly small chips, so costs per determination really do drop. One thing it's increasingly used to look for is single base pair differences, known as SNPs, snips. It was $1 per snip and went to a dime. It's probably on its way to well under a penny per snip. Pretty impressive. Affymetrix uses photomasks like the chip guys, but there's nothing electrical about their product."

"Nothing?"

"The chip is not electronic. Again, think biology. You have to look at the input and the output. The first thing you do is extract DNA or RNA from the sample you want to analyze and then you amplify it and label it. The label makes it easy to read in the scanner, and you can use different colors of labels so you can analyze, for example, what differences exist between a cancer sample and a normal sample. The thing lights up like a Christmas tree, different colors in different locations, lets you know the underlying nucleic acid sequences found. Pretty complicated chemistry. It's a complex recipe—lots of steps, lots of time—you incubate it for 24 hours."

"It takes an entire day?" I asked.

"Welcome to biology."

"Jeez," I sighed.

"The vision, the holy grail, is known as 'Bleed to Read.' "

"I like that, bleed to read," I said. I'll have to steal that line.

"A pinprick on your finger, a drop of blood on some cheap, foolproof device and an answer seconds later—healthy, sick, check engine, whatever.

"With DNA, and most other types of testing, we're not there yet. You wouldn't believe all the stuff that is in your blood that isn't relevant to a diagnostic test—you've got to get rid of it for the test to work at all. Find the needle in the proverbial haystack. The expression is 'fractionate'—reduce the complexity until you have something you can actually measure. Prepare the blood sample, extract and amplify the DNA or RNA. It's not easy, this stuff is fragile. It's not a Quaker Instant Oatmeal, just zap-in-the-microwave process. It takes days. Even with robotics, it can still take many, many hours. A GeneChip fits in your hand, but a full system to come up with the inputs to the GeneChip and read the output would probably fill a kitchen counter. And you'd need Emeril Lagasse to operate it."

"So I should look elsewhere?" I wondered.

"It could scale. That's what I do," Will said.

"At Caliper?" I asked.

"We have this really cool liquid handling robotics, even our own biochip. We call it LabChip and it's basically microfluidics—channels the size of a human hair. Remove lots of the stuff you don't want ahead of the actual diagnostic test. It's like my super automatic espresso machine. It removes all the skill from making the perfect espresso every morning. My four-year-old can do it. It makes it ready for broad deployment. Even doctors can figure it out."

"So DNA sequencing and all that stuff is going to be reality. I can sequence my genes and get my snips on the cheap?"

"Yeah, I guess that's possible somewhere out there. But you know, as important as DNA is, in many cases, it's only a clue, not the answer."

I poured Will a full glass of IPA. This is what I was waiting to hear about.

"What do you mean?" I asked. Will got a serious look on his face. I'm not sure I've ever seen that.

"Mapping the human genome was billed as a Nobel Prize–winning breakthrough, right? The 'digitalization of life' "—Will threw up air quotes with his fingers—"that would enable complete understanding of disease, and in the process make savvy investors wealthy. Yeah, right."

"It did for a few minutes in 1999," I prompted.

"Look, that's all good, but mapping the human genome

was a starting point—it raised more questions than it answered. The main question when it was done was simply 'what do we do now?' Maybe it's because no one actually, really believed that it could be done so quickly."

"No one thought it out?" I asked. Will ignored me.

"You know I used to work for Applera. In 1998, we hired 'noted DNA genius' "—air quotes again—"Craig Venter and put him in charge of Celera Genomics—which was just a tracking stock of Applera."

I nodded.

"We gave him early access to some 200 next-generation DNA sequencers and all the capital he needed to build a world-beating sequencing factory. That, along with his novel 'shotgun sequencing' strategy and fuck-you attitude toward the scientific hierarchy, made history. There's been zillions of books written about Celera, you probably know the story. What's probably not known is that a couple of us were trying to get Celera to position themselves to be viable in the postgenome world. No profits, less than $100 million in revenues, but a $14 billion market cap. Funny what kind of a mood that will put you in, for a brief moment, anyway. All I know is one minute we're in Craig's office watching a video of a 'rock star' public relations TV appearance he had just made and he's joking about the Porsche dealership that is going to open in the Celera lobby to service all the option-rich employees, and then, boom, the lights flick on and the party's over in more ways than one."

I refilled his glass.

"So Celera was on top of the world and then complete disaster struck. There's this carefully crafted and negotiated conclusion to a noisy race, both Celera and the public Human Genome Project finished the effort on the same day—thank you, Bill Clinton. But that wasn't the disaster. The meltdown came a couple of days later when Celera woke up with a hangover, realized the human genome was sequenced and couldn't figure out for the life of them how to make a living. Within a year, Venter was gone. You know why?"

"Nope," I said. No way was I getting in front of this freight train. Let it roll.

"Wasn't Venter 100% successful, beating all odds? Didn't he do exactly what he promised, even years ahead of schedule? Sure, but the sad thing is that the human genome wasn't really much of an answer to anything. It became a commodity overnight and hundreds of millions plus of additional dollars needed to be invested to give the company a chance at figuring out how to get any direct value, meaning drugs or diagnostics, out of it. It was a rat race—the mouse on the wheel who doesn't know how to get off, but the cost to stay on the wheel keeps going up."

"You're mixing rats and mice, but go on," I noted.

"Reminds me of my favorite conversation with Craig. He is a hard-core sailor. I saw his boat, *Sorcerer*, at Nantucket one summer, racing in the Nantucket Bucket, which is a race series for sailboats 75 to 175 feet in length. The following fall, I asked him about his boat. He told me he sold it. Too expensive to keep. Needed three full-time crew. Didn't get to use it enough. I mourned his loss and asked, 'So, you are getting out of sailing?' He looked at me funny. 'Of course not, I'm buying a bigger boat.' The guy can't get off the mouse wheel. I'm sure he knows that he's got to run even faster to come up with anything really important."

"But the DNA sequence isn't worthless, is it?" I asked.

"No way. But to extract value from the genome requires an analog expertise, drug discovery processes, diagnostics, whatever . . . not Craig's digital DNA expertise. Craig moved on and Celera is now an analog company, still leveraging DNA, but it's just like every other biotech wannabe—chemistry, screening and the like, even though it's the postgenomics world."

"So biology just stays analog?" I was mixing my own metaphors.

"Nope, it will all become digital over time. It has to. But it's messy analog—lots of interference and a few crossed wires here and there. The genome, the transcriptome, the peptidome, the proteome, the metabolome are all efforts to digitize understanding of analog biology. It's not so easy. It's really systems biology that matters. You can digitize some of it, but still not understand the entire system—like how is it possible that there are over one million proteins in the human body but less than 30,000 genes?"

"Uh?" I sputtered.

"The answer is that you have to move beyond DNA. DNA is just a blueprint for a house, for example. You might have a bad blueprint, which would be an important problem. But did the construction company properly execute against the blueprint? And what about the tenants—are they destroying the place or taking care of it? Is the trash piling up or being taken out routinely? Is the microwave interfering with the Wi-Fi? Is . . . ? "

"Okay, I get the point." I said. "What you're saying is if something's broken, I need to look at the house, not the blueprints."

"Bingo: 30,000 instructions are the starting point. People smarter than me can tell you about expression, translation, gene splicing, posttranslational modifications, whatever. But at the end of the day, it's a million or more proteins that make the body function. DNA is important, and sometimes even the answer, but—"

Will took a deep breath. I think he was about to get off the soapbox.

"But it's nasty complex. You need what's known as a surrogate end point outcome. A smoke signal that something is wrong before you've got some big fucking hairy tumor killing you. In fact people try to find these biological markers, some protein that acts as a smoke signal. If there is smoke coming out of the window of your house, there's probably a fire somewhere and—"

"Okay, okay. No more house stuff. Maybe it's your automated espresso machine that blew up," I protested.

"So that's it. There's no simple answer for disease. The 'one gene, one protein' dogma has been replaced with something a lot more complicated and a lot less understood. It will need digital technology, stuff that scales. Then it will settle out. Technology will digest it and reduce it to digital, scalable, cost-effective, change-your-world, change-your-life science. Push-button espresso. No flames. We're just all impatient."

I got the bill. Cheapest education I'd ever received.

Chapter 28

■

Who Pays?

After watching heart scans and virtual colonoscopies and thinking out all sorts of neat stuff that might be possible, I always ended up with the same question: Who's going to pay for all this stuff?

Blue Cross had already turned me down. Didn't they care about me?

I'd always thought insurance was a scam (until I totaled a car and got a check within two weeks). But health and life insurance were always confusing to me—they seem to have things backward.

Think about it: you pay premiums year after year to a life insurance company, which will pay a lump sum to your heirs if you die. Life insurance companies would like to keep you alive forever.

On the flip side, health insurance companies pay your medical bills. If you get catastrophically sick, the incentives suggest your health care company would probably rather see you dead.

I'd like my life insurance company to pay my medical bills and my health insurance company to pay a huge fee if I die young. Now *that* would be incentive.

But back to reality . . .

Going into the 1980s, health care costs were soaring and the insurance payers didn't like it one bit. Either they raised premiums to oppressive levels or found another solution. A movement called managed care came out of the land of fruits and nuts in California and slowly made its way across the U.S. HMOs, co-pays, restrictions on procedures, no choice of physicians, made a run but ultimately was rejected by consumers who preferred choices. Today, it seems, we have a hybrid.

Still, to contain costs, payers have adopted Medicare's DRGs, those Diagnostic Related Groups, to help limit payments. They also use something called CPT and ID-9 codes to fix the reimbursement on various procedures.

Every possible procedure is spelled out in excruciating detail at the Department of Health and Human Services Center for Medicare and Medicaid Services, with prices set by some bureaucrat who probably flunked out of community college. How much to reimburse for American Medical Systems' Acticon Neosphincter? I'm glad you asked. Just use code ID-9/49.75 Implantation of Artificial Anal Sphincter or even better, ID-9/49.76 Removal of Artificial Anal Sphincter.

Unless you can bill patients directly, the only way for doctors to get paid is to have a code. Nuclear imaging 78461—a cool $311.82. An echocardiogram is DPT 93307—$411.83 all-in, ding-ding-ding. That DRG 125 in the cath lab—$5,299, ka-ching!

I'm not suggesting, too strongly anyway, that doctors are greedy bastards who prescribe tests all day long and then perform them for fat fees. But I am suggesting that the incentive structure is there and doctors can easily hide behind their Hippocratic oath and malpractice premiums and run and get reimbursed for tests galore. Are they good tests, bad tests, inconclusive tests? Well, if it picks up one diseased heart out of 100, then it's worth it, right?

Someone told me the Mayo Clinic does 75,000 echocardiograms a year. Nice business if you can get it. Necessary? Perhaps.

In the emergency room of lots of hospitals, you are likely to find a CT scanner. Just about everyone who comes into the ER with a headache or bloodshot eyes gets a scan, with a nice charge of $500 for the hospital. Is it needed? Debatable.

There is plenty of talk of a single-payer system. No insurance needed, the government pays for health care. But how do you combat overusage? Today, in Canada and the U.K., rationing solves this problem. That doesn't seem like the American way. But when a bureaucrat in D.C. is setting prices for neosphincters, it doesn't sound like we are all that far from single payer.

Others suggest Health Savings Accounts, or HSAs, are the answer. Patients with pretax dollars and, say, a $1,000 deductible, will be careful how their personal health care dollars are spent. I used to agree, until someone pointed out that as soon as you walk through the door of an emergency room, you'd blow through that $1,000 deductible in the first five minutes. HSAs go only so far.

Can we afford all this newfangled technology in health care? Good question. The way the current system is going with the perverse incentives for self-referral (doctors referring patients to their own facilities), it's not clear if new technology will actually save money. I tripped across an article by a Dr. Laurence Baker at Stanford University, which states quite clearly that new technology and increased medical spending are interrelated. You invent a new scanner and medical spending goes up. A quick email asked Dr. Baker if we could meet. The next morning, we were sipping coffee on a 75-degree sunny day.

"You know about cholecystectomies?" Dr. Baker asked.

"Uh, no." This wasn't starting well.

"Gallbladder laparoscopic surgery. Minimally invasive.

You cut a small slit and with, well, chopsticks, perform an ablation of the gallbladder. We used to cut the patient open, requiring a two-week, $20,000 hospital stay. Now, it's a same day, 30-minute procedure, and a tenth of the cost because there's no hospital stay. That's the key." Sounds like Silicon Valley scale to me!

"So, that's good?" I asked.

"Well, sort of. Procedures are up 300% and spending on gallbladders is up 50%."

"That's bad?" I asked. I'm a great straight man.

"Health is better, but costs are not down. New technology drives more spending. It's just hard to quantify better health."

"Okay, I'm still confused. How do you quantify it?" I asked.

"By law, Medicare is not supposed to sort things by cost, but it's inevitable. The dirty little secret in the industry is something called dollars per life-year. Some vaccines are expensive but extend life. They may cost $1,000 per year, but that seems acceptable. ICDs . . ." I must have looked confused. "Uh, implantable cardioverter defibrillators, run $20,000 but can extend life by 5, 10, 20 years, Medicare now pays for them."

Now we are getting somewhere. I can envision a spreadsheet and . . .

"But don't try to do real quantification." Dr. Baker interrupted my thoughts. "I do this for a living and have a hard time. There's not enough real data. Medical technology companies don't want data out there, just in case it's a high dollar amount per life-year. I do this assessment, or try to do cost-benefit analysis. It's hard."

"But it has to be done or spending just goes through the roof?" I asked.

"Sure. Imaging is a good place to start. Except for some radiation, it's passive and doesn't harm patients. We need to get it out there and see if it is worth it."

"Why don't we?" I asked.

"It's all about the codes. Doctors do what they get paid for." This sounds familiar.

"The problem is overregulation of the front end," Dr. Baker continued. "We need to let new technology proliferate some, learn by doing. If we dramatically restrict use of early technologies, we won't have the natural evolution."

"And get stuck with echoes and nuclear imaging?" It seemed odd that even though all this great new stuff exists, doctors continue to use old tools.

"Right. We need experimental codes, the power of approval for doctors to get paid for new systems, and then generate data to see if it's worthwhile, not just take GE's word for it. Think of it like Phase III trials for new drugs—broad usage with data feedback. The dollar per life-year is easy to figure if we experiment. Frankly, consumers will be the ones demanding these things."

It was starting to sound more like the Silicon Valley model. Try new things, see if customers find them productive, and then invent new ones to replace the old ones.

It's that fuzzy black-and-white analog world in need of *Wizard of Oz* color and digitization. Once you go digital, you can never go back and you slide down that slippery learning curve of constantly decreasing prices.

But you can almost tell that we will spend more, not less. At least at first. I'd pay more for early detection, but my guess is there is a great spreadsheet in the sky that calculates at what cost it is cheaper to check for problems early than pay for treatment later. That would be an interesting formula to see.

You just know that when these new technologies take hold, wider usage will drive cost per unit down. Just like 3-D graphics cards and digital cameras and everything else. We've seen this movie before.

It's economics—but as with the cholesterol conspiracy, I

suspect there are plenty more hidden vested interests that muck it all up.

Nope. Zero-sum economics, a damn spreadsheet alone, won't cut it. The answer is somewhere else—outside of this rather large box. Since the expense is in the catastrophic end of the curve—heart, stroke and cancer, the Big Three—the real answer's got to be there somewhere.

Chapter 29

■

Can Biotech Deliver?

I drove around three times looking for Building 4. This company was bigger than I thought. I hadn't even heard of them until that morning—yet another biotech firm in South San Francisco. Dime a dozen. I was running late and finally found the main lobby and begged for directions. I was directed back the way I had just driven to find Building 4. "We don't label them for security reasons," I was told. Great. Here's a company with $50 million in sales dropping $100 million a year in losses. Who would miss them?

"So how do you come up with new drugs?" someone asked as I snuck in and found a seat in the back row.

"That's a great question," the CEO replied. I've learned over the years that you know you are in for a long-winded and mostly worthless answer when someone gives the "great question" preamble.

"Let me give you an example. We've found in preclinical studies that one of our compounds, number 184, has shown potent inhibition of the hepatocyte growth factor receptor and vascular endothelial growth factor receptor 2 . . ." Blah, blah, blah. I was lost. ". . . and demonstrated significant oral bioavailability and excellent pharmacokinetic properties. This is very exciting stuff."

I'm sure it is—whatever it is. I suddenly felt depressed. I had done a lot of work already and it turned out I still didn't know a goddamn thing. I still had a lot to learn.

"But how did you find this compound 184?" someone asked, which is what I wanted to know as well.

He babbled on for about five minutes, and despite trying to focus as hard as I could, I caught only a few snippets that I could sink my teeth into—something about running the target against a compound library; exhaustive tests looking for reactive levels, or something.

I lobbed in a question from the back row. "I guess I still don't understand how you find potential drugs. It sounds to me like you throw spaghetti against the wall and then you take what drips down and test it against tumors to see if they shrink."

"Well, I hadn't heard it quite that way, but you aren't that far off," answered the CEO. "We get targets from pieces of tumor or other cancer localities. We then introduce recombinant versions of this target into tiny reactive chambers filled with potential compounds—and see if anything happens. If something does, then we have an experimental compound to work with."

"How many compounds?" I asked.

"Our proprietary library is now up to 4 million compounds."

"Like what?" someone else asked.

I wouldn't have been surprised if he answered lizard lint, pixie dust and an eye of newt. He more or less did.

"We scour the globe for invertebrate and vertebrate artifacts that we can use to create small molecule compounds that might be highly reactive against our target, er, targets."

"And you run tests?" someone else asked.

"Let me take you to our laboratories and you can see for yourself."

The group followed the CEO out of the conference room, outdoors into Building 7 or 8—who knew, it wasn't labeled.

We were ushered into a rather warm room with a giant robot arm in the middle and stacks and stacks of—were those miniature ice cube trays? The robot was picking up the trays and shoving them into what looked like a Susie Homemaker toy oven, batteries not included. Who knew what happened in there.

"As you can imagine, testing a target against 4 million compounds would be prohibitively labor intensive, so one of our proprietary tools is the automation of the screening process. Each receptor assay"—I think he was talking about the ice cube trays—"contains 512 compounds in a micro reagent cell. This robot loads the assays into testing chambers 24 hours a day, where the target is introduced and our screening begins. We can then test each cell for some sort of change and narrow down compounds that might be worth focusing on."

This is it? Spaghetti thrown against the wall in the hope that one or two strands would stick? Crazy.

The CEO quickly moved us into another lab, which looked like a high school chemistry class—beakers, Bunsen burners and all.

"Once compounds are identified, we start manipulating them, figuring out the chemical structure of active compounds, and then using discovery pharmacology and informatics to try to marry the compound to the target, if you will."

"Huh?"

"It's our version of rational drug design," he said, sensing the confusion.

"But isn't rational drug design a top-down approach, you

create drugs from scratch based on the structure of the targets," the guy next to me on the tour wanted to know.

"Well, sure, but that probably doesn't really work." Didn't that Class of '55 dude tell me the same thing? "We do the screen first, then dig in and figure out the structures to see if they might be compatible. In the next room, we have a 3-D lab, where combinatorial chemists and others can view actual protein structures using glasses. We'll equip everyone with the latest liquid crystal shutter imagers. Every 60th of a second, our computers calculate what to show your left eye while blocking out the right . . ."

I'd heard this before. It was my turn to start fidgeting. If I understood this right, these biotech companies were nothing more than giant sniffers, screeners that looked at everything and anything that might actually have some interaction with cancer cells. That sounded more like trying to buy up every Willy Wonka chocolate bar to find the golden ticket. Where was the science in that? It was a giant crapshoot with shareholders' money. Worse, where was the technology that scaled? Robots didn't scale. Chemistry didn't scale, as far as I could tell so far.

We sat through the 3-D demo on a five-foot diagonal screen. Not even close to as cool as the CAVE. Molecular models, protein folds. I pulled one presenter aside after the lights went up and everyone was shuffling out and asked if they really used this stuff or if it's just what is affectionately known in Silicon Valley as demoware. He got a funny grin on his face and said, "Put your glasses back in the box and we'll move onto the next part of the tour."

"Once a compound is found to have any positive reactive results, it's just the start of a long journey." This was the understatement of the pharmaceutical industry's problems.

"Figure five to seven years. If we found something while

you guys were in with the robots, and I can check in the morning to see if anything interesting came up, it's probably a billion dollars to get it to market. You don't need a scientific calculator to figure that even at a lofty $1,000 a month for the drugs over five-year therapy, we'd need a few tens of thousands of patients just to break even.

"Then we have to determine the small molecule–large molecule thing. A small molecule drug can pass through the gastric walls quickly without breaking down, so it can be offered orally, in pill form. Large molecule drugs can't get across cell membranes, and as proteins, are metabolized the same way the protein in a burger is, and therefore need to be injected instead of taken in pill form, which drops the available market by a factor of 10, maybe 100.

"We've got bioanalysis during preclinical, then toxico-kinetic, pharmokinetic screening. Add bioequivalency, drug interaction studies, immunoanalytical, it goes on and on. And that's before we get to in vitro efficacy tests during the various phase trials. The FDA sets the hurdle fairly high."

Someone lobbed in, "What happens to the potential drugs that don't work?" What a stupid question.

"We have closets filled with compounds that looked promising, had positive reactions, even modest efficacy in limited trials, but never really had a chance to come to market. Maybe they needed higher doses that had too much toxicity. They clearly worked on certain tumors, maybe even started to shrink them, but had some side effects, so they got shelved. It's disheartening when you get close, for the patients, of course, and for us, as the spending comes to naught. We study our losses, but mostly we just move on, find new compounds to work on."

From the back of the room, someone lobbed in a question that I immediately wished I had asked. "I notice you guys only have $150 million in cash and are losing $100 million a year.

How can you afford the $800 million for even one potential drug?"

"Great question," the CEO answered, buying himself time to come up with some words that would put the best light on his company's dire predicament.

"We are really just the farm team . . . get it, P-H-A-R-M team. Ha, no, well, okay. The way the industry is structured is you have these Big Pharma guys with lots of money but not many interesting drugs in their pipeline. So they pay us for distribution rights of our promising compounds. It covers our research and development costs and gives us a deep-pocketed partner to market and sell our successes."

"But don't you give up most of the—" someone asked.

"It's a bit like producing movies, you put in all the sweat and toil and hope it's a blockbuster so you can share in its success."

"Is it as crooked as Hollywood accounting?" someone else asked. Lots of laughter. Damn, I wish I had asked that one, too.

"So let's move on to the final stop on our tour, in Building 10, the Wall of Fame, testimonials from famous scientists around the globe as to our . . ."

This guy had 18 months to live and was rather upbeat about it. His cash was draining fast and his robots were working overtime for some compound that would cure some subset of colorectal cancer. He either had to raise more money from the (unsuspecting) public or get gouged by a big pharmaceutical company and cut a Faustian bargain to stay alive, giving up future sales for piddling royalties. "You need a deal real bad? Well, here's a really bad deal." Hell of a way to make a living.

Okay, if it's not spaghetti against the wall, just trying compound after compound until there is some reaction, are there other ways to find new drugs?

Rational drug design has been talked about over and over and somewhat discredited. Despite a flashy book named *Billion-Dollar Molecule* about Vertex Pharmaceuticals in Massachusetts, there just aren't that many drugs that are designed using 3-D glasses and getting proteins to fit the cells of targets just perfectly. That grumpy old guy in the CAVE at Cornell may have been right. Or maybe it's just too early.

Is there some semirational way to create drugs?

Chapter 30

∎

Monoclonal Antibodies

High school physics teaches us that every action has an equal and opposite reaction. So too, every protein has an equal and opposite antibody—actually, more than one, they react against different parts of the protein's surface. That's how our immune system works, or is supposed to work, anyway. Bacteria, viruses, cancer cells, all known as antigens. Each has proteins that can be attacked with antibodies (which are also proteins, but don't get confused, this isn't rocket science).

Sometimes it takes getting a mild form of a disease to build up antibodies to fight a nasty form. Think smallpox. The vaccination involves getting relatively mild cowpox, and besides a scar on your arm, you have acquired the antibodies to fight smallpox, so occurrences of that disease are rare.

Each cancer has unique proteins associated with it. Can't scientists just create antibodies for every known cancer protein and be done with them?

Not so easy. But it's a multibillion-dollar effort to try.

No one has volunteered to have cancer cells implanted in them, just so researchers could harvest antibodies. Ethics. Damn.

But when there's a will, there's a way. Fortunately, mice have no natural political constituency. Hundreds of thousands, maybe millions of genetically engineered mice have stepped forward to save humanity. Rent the movie *Willard* this weekend and be thankful.

One drug design effort is from a company called Protein Design Labs. I should start out by noting that PDL, despite being worth $3 billion, has lost money for 19 straight years and blown through some $270 million in capital.

Their mission is fairly straightforward: create antibodies for humans from information gleaned from mice. Antibodies are proteins that our immune systems produce to fight anything foreign in our bodies that could kill us—viruses, bacteria, toxins and even other proteins. Our bodies produce them naturally, unless of course they don't. Some gene is missing or corrupted or some other internal function is kaput.

The easiest way to create antibodies would be to inject these substances and proteins one at a time into somebody and then extract the antibodies that are produced and give them to someone who needs them. Call it the Mengele technique because it is illegal, unethical and more than creepy.

But mice will do. They reproduce like rabbits, don't take up much space and don't have a hell of a lot else to do. Sure, they all want to grow up to be rats and live in the New York subway system, but don't we all.

Inject cancer cells or just specific proteins into mice, and they quickly create antibodies to fight the proteins they find. These antibodies can be harvested from the blood serum, but still aren't quite ready for prime time. They are not terribly pure, and worse, they are mouse antibodies, not human antibodies. Our bodies know the difference and fight them with our own defenses.

PDL injects various stuff into mice and then pulls out the antibodies produced. PDL plays around with them, fuses some of the mouse antibody amino acids with human antibody amino acids and watches what happens. Every once in a while, you get something that just might be effective as a drug or linked to other substances to make them more effective. PDL hopes to bring out one new drug per year to clinical trials. Think about that, $270 million invested over 19 years and they can hope, pray for one new drug per year. It's not easy. Glad to see that mice are of some use. I've had some funny run-ins with them.

One problem is that these mouse or even human antibodies don't get created in massive volume, just enough for the immune system to fight off danger.

Back in 1975, two scientists, Georges Kohler and Cesar Milstein, attacked this problem. They fused two different cells together, antibody-producing cells and myeloma cells that are considered "immortal" (yup, the bone marrow cancer ones). What they ended up with were antibody-producing cells that were immortal. The antibodies produced are called monoclonal, since they come from just one type of cell and are quite pure. Think of it as a factory in a flask, cranking out antibodies.

This was Nobel Prize–winning stuff—they took the 1984 Nobel prize for this "hybridoma." Using cancer to fight cancer has a certain symmetry to it. The work immediately ushered in the world of miracle cures, but strangely, for a while, not much really happened.

In the late 80s, this technique was resurrected as yet another method for drug discovery. Does it work? Kinda, sorta, maybe, who knows? Genentech has spent billions on monoclonal antibody drugs. Acastin fights colon cancer, some of the time, anyway. Lots of doctors prescribe it, off label, for breast and lung cancer. Herceptin is another monoclonal antibody, which is effective for about 20% to 30% of breast tumors.

Why? Well, it's an antibody. Only if the tumor expresses a certain growth factor receptor, known as HER2, and relies on it for growth, can Herceptin be effective.

Martha Stewart's portfolio's favorite drug, Erbitux from ImClone, is an MAb that inhibits epidermal growth factor, a protein that stimulates certain types of cell growth. The FDA finally approved it, after lots of dramatics, for colorectal cancer. Does it work? Sometimes. Same goes for Iressa, another cancer drug that looks for cells with a certain mutant epidermal growth factor receptor protein—if no mutation, Iressa is helpless.

MAbs as drugs are promising, but no cigar.

But the cool thing about monoclonal antibodies is that they attach themselves to very specific proteins, so they are tiny spies and are increasingly used in diagnostics.

Those throwaway pregnancy test kits that cause fainting spells in women and fright flight impulses in an equal or greater number of men worldwide are nothing more than monoclonal antibodies, MAbs, finding specific proteins in urine and turning blue in the shape of a plus sign (how do they do that?).

The technique is called an ELISA, enzyme linked immunosorbent assay, and came about in the early 70s. In an ELISA, the MAb is actually an antigen, chemically cemented to a surface. Then a polyclonal antibody, with a color-changing enzyme attached to it, finds and binds to the MAbs. When urine or blood with specific proteins flys by, the whole solution changes color. Pretty simple to use, but not cheap to make. It's probably 12 to 18 months, lots of dead mice with harvested spleens, and $100,000. The test kits can run $10, $20, or more. But they work.

I wonder if the end game is mice as pets/antibody factories. Yuck.

Part IV

■

Chapter 31

Snow Pea Diet

I was exhausted. I had convinced my wife that skiing would save my life, but a combination of the high altitude and skiing like a maniac since nine in the morning meant I had zero energy left in my body. Luckily, my friend Mark Balestra had taken my advice from my last trip when I gently explained to him that he was a complete idiot and that a ski house without a hot tub is like a joke without a punch line—completely unacceptable.

So I find myself sitting in a brand-new hot tub at his ski house in Utah during a much-deserved boys' weekend of skiing. How does that expression go—I can't use any more friends because I'm not done using the ones I already have.

Even better, Mark's friend, who goes by the name Nuts—I was told not to ask—had stocked the snow around the hot tub with long-neck Polygamy Porters—life was good. I was as happy as a pig in warm shit.

As was my usual MO, I was going on my second hour in the tub and had been visited by everyone on the trip, in 10- to 15-minute clips. I noticed that another guy, Robert Gaffney, was also putting in serious time in the tub. It was good to have company and someone to revive me if I passed out from the heat. He was a neighbor who I had met a few times at cocktail parties, but I didn't know him all that well. We were talking about kids and schools, the usual stuff, when Nuts came out and asked if we wanted to go out for dinner or bring in takeout. My legs had turned into some rubbery substance and I was in no mood to leave the premises until the beers ran out. Robert agreed.

"So what do you have a taste for?" Nuts asked. "Pizza, burgers, Chinese, not sure what else they have up here, the pickings might be a little slim."

"No Chinese," Robert quickly threw in.

"You sure?" Nuts asked. "Seems like a few takers inside."

"Chinese is okay, I guess," Robert conceded. "Just no snow peas."

"Really? You allergic?" I asked.

"Nope, just can't have them," Robert said, shaking his head.

"Let's get pizza," I said. "One cheese, one with pep, and one with double snow peas," I said.

"Whatever. I'll bring more beers."

"So what's with the snow peas? I can't have Wild Turkey or Green Chartreuse anymore, but I'm not alone on those."

"Well, we don't know each other that well," Robert said looking around to see if anyone else was listening. It felt like some deep dark family secret was about to be revealed. "Do you know what I do?"

I had no clue. "I have no clue," I told him. I started thinking food tester or Chinese immigrant smuggler or vampire slayer or grave robber. Gotta lay off the beers at high altitude.

"I work for one of these medical device companies in Palo Alto," he told me.

"Oh," I said.

"Yeah, it's one of those long-haul things, we think we have a procedure to help with obesity."

"Cool." I always love hearing what other people do, but usually don't ask, because it often means an hour or more of tedious trivia about two-tier enterprise software sales channels. This might get interesting, plus it was too cold outside to get out of the hot tub.

"It's one of those huge markets, pun both intended and unintended. There are 5 or 10 million Americans who are morbidly obese, but another 60-plus million who are just plain old obese."

I pulled in my gut, even though it was underwater.

"Diets are a joke, drugs don't work, binge and purge is not socially acceptable, so it comes down to medical procedures. For the morbidly obese, that usually means a gastric bypass operation."

"So?" I asked.

"So, it costs 30 grand, they cut you open, and there can be complications. But there are plenty of folks who are desperate—and the risk from an invasive operation is nothing compared to their risks from diabetes and heart attacks, the kind of stuff that goes along with obesity. Only 1% actually have a procedure done."

"That's all? I would have thought it was more," I said.

"That's it. It's not easy. It's called roux-en-Y. You go in, staple the stomach so there are two pieces, a small one on top that the food goes into while the bottom one still produces acid and, well, stomach stuff. But most of the stomach and small intestines are bypassed, a hole is cut in that upper stomach piece, and a connection is made directly to the large intestines. It's a double whammy. A smaller stomach so you feel full faster, and less digestive surface area to absorb the food, in case the patient eats through the procedure."

"Eats through?" I asked.

"Yeah, they found most staple stomach procedures didn't

work because the patient would just eat all day, milk shakes, peanut butter cups. They'd fight off that full feeling, so you had to do a bypass to slow absorption."

"Can't all, er, porky people get this?" I asked.

"Nah, too expensive and too radical. So now they have Lap Bands, which go around the outside of your stomach and turn it into a sort of an hourglass, only a little sand can get through at one time, so you feel full. They can even adjust the bands over time. But it's still an invasive procedure. So is VBG, vertical banded gastroplasty, which creates a separate pouch. They still have to cut you, so there is recovery time. And it ain't cheap."

"So what are you working on, a bionic stomach that rejects Reese's products?"

"The idea is if doctors don't have to cut you open, a procedure is both cheaper and potentially less dangerous. We figure if we can get it to 5,000 bucks, from 30 or 40 grand, there is a whole new market."

"Uh-huh." I resisted more stomach jokes because I was intrigued. I was also close to passing out from overtubbing. I yelled for Nuts to open two more porters for us.

"We go in like a GI. You know, a gastrointestinal probe. We just go down through the throat and do what we need to. We send down a tube with a vacuum pod. It opens up when it gets there and pulls parts of the stomach together that we can staple from the inside. You are left with a small pouch in your stomach. When you eat, the pouch fills up, and your lower esophageal sphincter closes and tells your brain that you are full."

I wondered if that was like that artificial sphincter I had read about.

"Pretty simple, actually. It's not foolproof, nothing is. It's probably not for the morbidly obese, but the market for merely obese is 10 to 20 times as large. This could be an outpatient operation, come by in the morning and leave for a nice, and smaller, lunch. $5,000, thank you very much. A couple of million of those a year and I stop working."

"Then you'll eat and drink all day and need one of these procedures."

"Bingo. What's really neat about this is the compounding effect," Robert explained.

"The what?" I asked.

"Well, if you are overweight or obese, let alone morbidly obese, chances are extremely high that you'll have some other complications. Most likely diabetes. But probably heart problems, back problems, liver problems, you name it, it's somewhat likely. So five grand for our procedure is pretty cheap if you can keep the patient from becoming a very expensive diabetes patient. At least that's how we are pitching it to insurance companies. We'll see if they agree."

I started thinking of all the doctors who treat obesity who might need something else to do.

"Get it to $500 and it's all over. So what's with the snow peas?"

"Oh, yeah. We're not ready for humans, yet. Live ones, anyway. We test our procedure on dogs, and cadavers."

"Yuck."

"Yeah, there is pretty much an endless supply of cadavers. When you go for a GI, your doctor tells you not to eat for 12 or 24 hours, but we really don't get a chance to instruct cadavers."

"Okay, I get it."

"The nerve of these people to have eaten before they dropped dead."

"Okay, you can stop."

"So I was scoping this guy, checking what things looked like before we started, and—"

I turned up the jets to high and started humming.

"And sure enough, there is this large, rather undigested snow pea sitting right—"

"Hmmmmmm, hmmmmmmmmm."

"—at the bottom of his stomach, so I asked my assistant, 'Does that look like a snow pea to you?' and I got back a 'Yup, a number five special—General Tso's Chicken with Snow Peas,' and anyway, no more snow peas for me."

Me neither.

The skiing the next day was great, and like my friend Brad Miller, I felt like a fucking rock star. Despite a few wipe-outs, yard sales and face-first tongue surfs, I avoided the infirmary. Maybe I've got a tumor somewhere, but I wasn't going to find out this trip.

But while hitting the bumps, I kept imagining rooms filled with cadavers. Spooky, right? I kept wondering: Why not do a CT scan on every cadaver? Then when you and I get our own live CT scans, doctors can compare our guts with those of millions of dead folks.

"Your arteries have no calcium, but your heart is very similar to those of three guys who all died of a heart attack."

Now, that is information I would find useful. And probably this, too:

"You know, Andy, your liver looks just like that of Tara Reid."

Fortunately, when I got home, images of dancing cadavers faded, to be replaced by thoughts of stomach stapling. I was still a few bags of Cheetos shy of needing help, but last I checked, the sidewalks were clogged with, shall we say, those who are gravity challenged.

I couldn't stop thinking back to Dr. Edward Manche, Doogie LASIK. The guy was raking it in by fixing eyeballs. He wasn't saving lives, unless you count Mr. Magoo not walking off a cliff. But it was almost a pure form of personalized medicine. Aesthetics. Vanity. Medicine to make us look and feel better—so much so that we are willing to pay for it out of our own pockets.

That LASIK eye surgery. Viagra. Stapling. Tummy tucks. Botox. Rogaine for male pattern baldness. Each of these in some form or another lowers costs.

The American Society of Plastic Surgeons, also known as the Nip and Tuck Yucks, lists the most popular procedures for 2004: 325,000 liposuctions, 305,000 nose jobs, 264,000 boob jobs (is that double counting?), 233,000 eyelid surgeries and a modest 114,000 face-lifts. Botox, although temporary, is certainly cheaper than complete facial reconstruction and skin transplants.

Maybe that's where medicine is really headed. We may get cancer, but damn, we'll look good.

What bothered me is that each of these was a one-off. But scale? I wasn't so sure.

Chapter 32

■

Angio

The Strikers were getting killed. The kids on the other team seemed a head taller—at least—and were roughing up our boys. I thought soccer was a noncontact sport, but apparently not for nine-year-olds.

I complained to Jeff Child, our team manager, who was standing on my right. "Those other kids taking steroids? We gotta get our kids on the juice, too."

"Seems like it. They are a year older," Jeff replied.

"Well, that's not right," I kidded. "We should probably start cheating."

A voice boomed from Jeff's right, "You know, Dan Reeves is a patient of mine."

It came from a 60-something-year-old guy who looked a little out of place at a CY Soccer game. I knew who Dan Reeves was—the old NFL coach of the Falcons and Giants and Broncos. The voice with a slight Texas accent had my attention.

"He once told me, 'If you can't beat 'em, cheat 'em a little bit.' "

A few chuckles came from the pack of parents watching the game.

" 'And if you *can* beat 'em, cheat 'em a little bit anyway,' " the voice added.

"That's good," Jeff said.

I whispered to Jeff, "Who is that guy?"

"Oh, that's John Simpson. His grandson is Jean-Claude, one of our backs."

"You mean the balloon guy?" I asked. I knew the name. He was Mr. Angioplasty.

"That's him," Jeff said.

Now it made sense. Dan Reeves had well-known heart problems and had undergone bypass surgery, if I remembered, missing the last two games of the Falcons 1998 season. Yet he somehow came back to coach the NFC Championship game and the Super Bowl, losing to his old Broncos team. You still see him on commercials hawking some cholesterol drug or another.

John Simpson, on the other hand, is a medical legend. Chances are his balloon catheter has saved the life of someone you know.

It was almost only Boston terriers.

Simpson planned on being a veterinarian but didn't quite have the grades. So he headed to Duke to get a Ph.D. in immunology, the study of the response of organisms to foreign substances. Which is kinda funny, since he ended up sticking a lot of foreign substances into bodies. He tossed the doctorate for an M.D. degree from Duke Medical School, and then moved to Stanford to study cardiology.

Just as he was finishing Stanford, he heard a talk by Andreas Gruentzig about, as John Simpson puts it, "this really neat procedure that he was working on where he would put a plastic balloon catheter into people's coronary arteries, blow it up, and it was going to make them a lot better. That seemed like a stretch—no pun intended."

Just about everyone was skeptical. As a cardiologist, Simpson wanted to remove the plaque. During one bypass operation, he accidentally dislodged a chunk of plaque and caused a nasty heart attack in his patient. It was at that point that he realized that what the balloon did was push the plaque back up to the surface of the artery. Plus, you could go in through an artery without cracking any ribs open. At the time the procedure was done through a patient's side, not through the groin.

He excitedly told his Stanford bosses about this great discovery and asked if he could go to Zurich that winter to study under Gruentzig. Their answer was basically, "Yeah, right. Winter in Zurich? Pay for it yourself." So he borrowed $500 and took off.

The Gruentzig system worked, but was really, really hard to use. Gruentzig had the ability to mentally map the patient's arteries in 3-D, a rare skill, especially in doctors. Nonetheless, this new field of angioplasty was revolutionary.

Back at Stanford, Simpson ordered a kit from Gruentzig to play around with, but it arrived without the balloons, which Gruentzig explained were in short supply. Ah, the mother of invention.

Simpson and a colleague, Ned Roberts, experimented with a bunch of different plastics, trying to come up with a balloon that would work as well as the ones Simpson saw in Zurich. They had three different balloons to try, and wangled time in a NASA animal lab at Stanford, and even a baboon to try them on. Unfortunately, they killed the baboon with the first balloon and weren't offered any more primates. So scramble they did, until they eventually found material for balloons that were easier to use.

Being in Silicon Valley in the late 70s, they did the only smart thing possible, which was to start a company, Advanced Cardiovascular Systems (ACS), to perfect the balloons and catheters. As with any good innovation, inventing it is not enough. It takes hundreds of iterations and lots of capital to turn the innovation into a real product. The beauty of the ACS system was an over-the-wire catheter that would find its way up the artery; balloons and, someday, stents would follow. Simpson and team eventually sold it to Guidant.

Simpson was never a big believer in stents, those little wire meshes that his balloons push out to help support the artery walls. I shouldn't say believer. They work. They have completely changed the way heart attacks are treated.

But Simpson's view is that the material that's in the coronary artery doesn't belong there. Don't leave the plaque there, let alone a wire mesh with the perfect nooks and crannies for more plaque to build up in. Get that stuff out.

Another Simpson company, Devices for Vascular Intervention, was put together do just that, to remove plaque, but it ended up with only modest commercial success and was eventually sold.

Too bad. Stents are amazing, certainly relative to cracking someone open and doing bypass surgery, but they're not without problems. Plaque can build again—restenosis. Newer drug-eluding stents, which are meant to help rebuild tissue on the walls of the stent, might cause blood clots, which defeats the purpose, doesn't it? So John Simpson is probably right: Get the plaque out!

I begged our soccer team water boy, Jeff Child, who mentioned he was on some board with John Simpson, for an introduction.

"Fine, but you're buying beers on our next visit to the Dutch Goose."

"Jeez. You drive a tough bargain. You buy your own deviled eggs and it's a deal."

"John, have you met Andy Kessler? His kid is Ryan, the one who just scored. He's doing some tech thing," Jeff said to John Simpson.

"Nice wheels on that kid."

"Thanks," I said. "His dad's not winning any races anytime soon. Listen, I'm working on a project on technology and health care and I wondered if I could buy you a cup of coffee sometime and . . ."

"Yeah, a cup of coffee is about as much time as it would take to tell you everything I know about technology and still leave time to talk about who's going to win the Super Bowl this year."

"Well, still, uh, thanks, then, okay. I'll be in touch." I went back to watching the game before I said anything stupid.

I got an email from Jeff later that day with Simpson's email and a two-word message: "Good luck."

Great. What do you ask the one guy who has seen it all and could rip my heart apart and put it back together with his eyes closed?

"Anything else new coming down the pike?"

Yeah, that wasn't going to work.

"Should I cut back on Egg McMuffins or do you think you'll be able to solve this whole heart attack thing?"

Okay, that's more like it. I couldn't quite ask it that way, but in effect, that's what I wanted to know.

> To: John Simpson
> From: Andy Kessler
> Subject: one question
>
> dr. simpson,
>
> we met on the soccer fields. that jean-claude is something else, he must have your genes.
>
> i have just one question for you. if imaging really does get cheap enough and we can find all this stuff that causes heart attacks and strokes, is there some device or procedure that will actually do preventative care, i.e. remove all that plaque before a heart attack or stroke?
>
> thanks,
> andy kessler

From: John Simpson
To: Andy Kessler
Subject: Re: one question

I am about to start a procedure but the short answer is
"YES" (and you should learn how to use capital letters).
John

I had this image of him dressed up in hospital garb, typing a
message with gloved thumbs into his BlackBerry as some poor
schmuck lay twitching on an operating table, and immediately
I felt bad that I had bothered him. On the other hand, if some
preventative procedure did exist, that poor schmuck would be
sipping piña coladas on his veranda instead of bothering Dr.
Simpson and holding up progress.

I finally caught up with him again. "Do you think early
detection will change things with heart disease?"

"The big limitation today with early detection of coronary artery disease is that angiography is the current definitive
method, but it's invasive." Dr. Simpson told me.

"And so—" I started.

"When a noninvasive method is shown to be as reliable as
angio—be it CT or MRI—that will be a big advance," he continued.

"And cheap enough?" I asked.

"When it's reliable, I think it will promote more aggressive treatment of early narrowings and patients will benefit dramatically."

Sounds like another way of saying, "If you're losing, cheat
'em—if you're winning, cheat 'em a little bit anyway."

I was not that surprised over his thoughts on reliability
over cost—I suppose that really is how medicine works. But no
doubt it would get cheaper.

He's got that covered. John Simpson's latest venture is a

public company named FoxHollow Technologies. It looks like a modern version of Devices for Vascular Intervention, a way to get the plaque out, not just push it up against the side of the arteries.

The SilverHawk Plaque Excision System sounds cool enough. Using the same minimally invasive catheters as do angioplasty and stents, it is more of a Roto-Rooter-like system to clear the arteries—snipping and sucking out hundreds of milligrams of plaque. Don't push it aside, get it out!

It hasn't completely proven itself. It has some application in Europe for cardiac and peripheral arteries, mainly legs. In the U.S., it's just peripheral so far. So John Simpson's quest continues.

I think if any of us had heart trouble, we'd like John Simpson to be in the operating room patching us up. In many ways, he already is.

So if high-resolution CT scanners can find blockage, will there be ways to get the plaque out? Simpson's Roto-Rooter is one of many attempts. An Israeli company named InSightec has a technology they call MRgFUS. I think that's Hebrew for melt it out. Actually, it stands for MR Focused Ultrasound. They use it now for the removal of uterine fibroids. You walk into a clinic, lie down in the machine, an MRI scanner locates your fibroids, and a beam of focused ultrasound energy is blasted just on the cells that need to be destroyed, raising the temperature in a very localized region. The MRI makes sure the beam isn't off course. Then you get up and go back to work after the ultimate outpatient procedure.

The company is doing trials to test the procedure for liver, breast and even brain tumors. From what I understand, arterial plaque is not far behind.

It's not hard to figure out that CT scans will be cheap. Dirt cheap. Doctors' office procedure cheap. $100 for a decent entry-level image of your heart and arteries. Enough to identify deadly plaque.

Why am I so confident? Because so much of the process uses silicon. Not the X-ray source, but the detectors and controllers and the storage and imaging software, and certainly the 3-D modeling and computer-aided detection software. Silicon means smaller, cheaper, faster, better.

But the next step is to do something about it. Something cheap enough to be proactive with arterial plaque—meaning a cheap procedure rather than a lifetime supply of Lipitor. When the two are combined, maybe Blue Cross of California and others will start paying for these heart scans. They're scared about the back end, paying for repairs. Once they know you have a life-threatening problem, doctors and insurers can't walk away from you. Hippocrats. So Blue Cross has the Sergeant Schultz defense—I see nothing, I know nothing—making them hypocrites.

But what's cheap enough? Angiograms are five grand and angioplasty with stents are some 15 grand, so you would think a couple of thousand dollars for a Roto-Rooter would make it an economically viable procedure. Or a few hundred bucks for a focused ultrasound blast. I think that's the lesson from imaging. It won't work and change medicine by itself. It needs the back end, the cleanup, to get cheaper, too.

Dr. Baker was right. The spreadsheets need to flash green. Not just finding but curing disease must be cheaper as a package before it rolls out in a big way. Until then, Pfizer and the conspiracy rule.

Chapter 33

■

256 Slice

There is a secretive arms race going on. This is the fastest technology race in health care, and I'm right in the middle of it."

I spoke to Dr. Jeffrey Goldman at Manhattan Diagnostic Radiology, on the Upper East Side, the sickly side of town if you measured it based on hospitals, doctors' offices and clinics. A radiologist, Dr. Goldman likes to push the state of the art.

"We got the first 64-slice scanner in New York. It's the Toshiba Aquilion 64. You know this model?" Dr. Goldman asked me.

He sounded like a techie bragging about his new PowerBook or Treo. My doctor brags about his new rubber hammer.

"It's great stuff—we can do a heart scan in not much more than 10 seconds, the whole procedure is 25 seconds, tops. The resolution is down to half a millimeter. What's different is that it's starting to break the barrier of conservative radiologists— they're actually starting to buy into this approach. Even better, we can do a 3-D model out the back end—some vessels are so small that slices just don't cut it anymore."

"So the industry will cut over to this stuff for screening?" I asked.

"Eventually. Look, this technology is very exciting at first, but it has real limitations."

Uh-oh.

"It's not that specific. Accuracy could be better. I can't really say what percent stenosis I am seeing. I've got nurses providing beta-blockers, measuring blood pressure, and I still get blurs. The heart is beating, and with lots of patients, it's beating irregularly, which is why they're here in the first place."

"So it's not for every patient?" I was disappointed—was my theory going out the window again?

"Not yet. But I'm starting to get hints of the next generation."

"Higher resolution?" I asked. Like high-end graphics cards.

"Well, we'll probably get that, but the next generation is about confidence, not resolution."

"I'm not sure what you mean." I said.

"With 256 slices we'll be able to scan the entire heart in 400 milliseconds, four tenths of a second. That's amazing—that's half a rotation. It's not the temporal resolution, it's that we can do it in less than one beat of the heart. No blurs. Plus we can check on things like perfusion—functional decreases in blood flow. This tool changes everything. We got a much clearer view of what's going on."

"So, smaller, faster, cheaper, better?" I quoted from the Silicon Valley bible.

"Smaller? Two-tenths of a millimeter would be amazing. Faster, yes. Better, definitely yes. But cheaper? I think out of the box they'll be $3 million. I suppose they'll get cheaper over time." Like all doctors, Dr. Goldman seemed to be thinking out loud. I wasn't going to stop him. "But then again, at 400 milliseconds, it's like a snapshot, I may not need beta-blockers. We can probably do twice as many patients each day, so cheaper? Yeah, we'll get there."

That's what I was hoping to hear.

"But it takes time."

Oops.

"You know about the overutilization problem? You know, self-referral?"

"I've heard about it," I said.

"Seems only cardiologists have this issue—echoes, stress tests. As a radiologist, I root for the imaging solution to hopefully get rid of unnecessary tests."

Does imaging make the problem worse? I was about to ask when Dr. Goldman blurted out, "There is the radiation question."

"Is it a problem?"

"Hard to say. In Japan and Europe they cut back on radiation in CT, so their images are fuzzier and not as useful. Maybe that defeats the purpose. 256-slice could have higher radiation doses. We don't know what the right number is. All the data is from Hiroshima survivors. Is that valid? It's an issue."

"So you think it can be a first-line screen?" I asked again.

"We do something like 1.2 million angios every year in the U.S. Do you cath or not cath? We get reimbursed $850 in New York for our scans, a CT is $500, going to $300. Reimbursements are going to drop as costs drop. A cath is two grand, at least. Today, it's really for someone with chest pains, those are the ones who should use this instead of a cath. Or an older patient with a bad stress test. Why go directly to cath when you can scan?"

He paused as he thought about his own question.

"I guess the bigger question is: When is this scanning good enough? There are lots of bad scans going on."

"I had a heart scan. Black-and-white slices, did I make a mistake."

"E-beam?"

"Yeah, EBT. Imatron, I think."

"Yeah, that's pretty dated. What did they tell you, a calcium score?"

"That's about it, mine was zero."

"Good for you. We can give you a calcium score, but also so much more. We can image cardiac function. That's invaluable information when fighting heart disease. Done right, used correctly I should say, imaging is ready for prime time."

"So I screwed up?"

"Not really." I was going to ask him if those older generation imagers could do really cheap scans, and if the industry should use them for the truly cheap first-line screen, then move up to better ones for the positive cases. But I couldn't get my question out fast enough.

"Then again, that all changes with 256-slice. 400 milliseconds is quick. Eventually it will be cheaper for me to provide. Yeah, I'd hold out for the best."

Chapter 34

■

Physical of the Future

Okay, I get it. This is the physical of the future. Sticking out your tongue and getting a finger up the wazoo is something some stooge might do, Dr. Howard, Dr. Fine and Dr. Howard back in the 1930s.

256-slice scanners, faster than your heartbeat, just might be the magic pill of diagnosis. It's as if doctors will be saying I was blind before I could see. Blind as a bat. Six blind doctors feeling around an elephant and describing a wall, spear, snake, tree, fan and a rope. Looking for clues in all the wrong places. Measuring cholesterol and blood pressure is like reading the outside temperature and humidity from inside your house and guessing if it's raining. Open the window, stick your goddamn hand outside and know for sure.

These scanners will spit out gigabytes of data, thousands of slices. We already know we can't afford radiologists to sift through all this data, not at their usual fees of $149 for 60 seconds.

But so what? Technology like R2 can help. It can flag anything suspicious. False positives will be a lot lower than the discredited scans from 2000. Higher resolution and higher speeds should mean artifacts go down, not up. Expertise embedded in silicon detector rows, 3-D rendering and computer-aided detection—not the Dr. Finkelsteins of the world. Again, not until we need him.

How much do these scans have to cost to become widespread? $500? $100? $20? It almost doesn't matter. The cost savings come over time. Spread the R&D over millions and you get scale. It works. But not yet.

I read there are upward of 2 million heart procedures done each year, about half in the U.S. Scanning 300 million Americans at even $100 a throw is $30 billion versus $15,000 for a stent times a million procedures or $15 billion. Today it's cheaper for your symptoms to show up in real life—clutching chest, can't breathe, life flashing in front of your eyes—than in a few gigs of a virtual scan.

Patients as consumers of medicine, well, that's a different story.

Let's see, I can buy that new Mustang or I can have a heart scan and not worry about dropping dead at work. Hmm. I have no doubt that lots of folks will take the Mustang behind door number two, but if we can spend billions on low-carb diets and Special K cereal and Egg Beaters, let alone Viagra and Botox, parting with some discretionary cash for a lifesaving heart scan doesn't seem so far-fetched.

No, this digital divide isn't going to be between the haves and have-nots. I suspect it will all get cheap enough, like PCs and cell phones. No, it's more the want-to-knows and the don't-want-to-knows.

If the end game of all this is more control of our health care, like managing your own money, some people are going to be good at it and others aren't. Maybe in some ways Darwinian—you become dependent on your own research and effort. It's going to take some combination of personality, character and intelligence just to be able to handle it—fight back the nausea when someone shows you your guts and glory.

I admit to sweating through my BVDs waiting for the results of my heart scan. I still see puppies being pulled out of a Boston terrier whenever I'm asked to confront anything medical. It's so, like, totally grodie.

But I suspect I just had to get over it. If I wanted to cheat a little bit, live a little longer than predetermined by my parental gene inheritance, I needed to wake up and smell the blood. Still, this is not a small obstacle to acceptance of this new medicine. Would you like to know when you're going to die? What if these new scanning guys threw a party and no one showed up?

But plenty will. The opposite question is, What if half of those million heart procedures are no longer necessary because something was detected early enough to treat in other ways? That's a lot of hospital beds and cardiologists and nurses with nothing to do. But that's exactly what is going to happen as scanning gets cheap enough, as Dr. Goldman's 256-slice machine gets utilized 24/7. He is the antihospital, antidoctor doctor.

Then, do we really need doctors—at least doctors as they exist today?

The geeks are definitely at the gate and chanting in strange tongues.

Chapter 35

■

Got My Blood Drawn

Okay, it wasn't easy, but I got finally got my blood drawn. I'm still stuck in the Three Stooges era of medicine. I filled up four test tubes' worth and FedExed the genetic kits off for testing.

I'm a little embarrassed to tell you how I got my blood drawn. Let me just say that this is what I didn't do:

Troll the back pages of *The Wave* magazine, one of those crappy local rags they give away for free in bars, chock full of ads for teeth whitening and car hot-rodding and oxygen bars and escort services, until I found an ad for a "Nurse to take care of your every need" and then ordered it up for our Friday night poker game, to everyone's delight—skimpy uniform, Red Cross nurse's hat—and then find out that the "actress" was actually a real nurse and would draw my blood, no problem, for the usual happy ending fee.

Really, I didn't do this.

I started getting nervous again. Not about the nurse, but about the genetic tests. What if I was some sort of mutant—had Godzilla genes for colon cancer. Did I really want to know? I thought so, but I was having second thoughts. Too late, I supposed.

By the way, these tests aren't cheap. Anxiety is expensive. $350 to test for a mutated MLH1 and MSH2 gene and the possibility of colon cancer. Another $400 bucks to test for a defective cystic fibrosis transmembrane regulatory (CFTR) gene. I think it's $400 bucks. They took my credit card. I'm not sure what amount is going to show up. Come to think of it, the same thing happened with the nurse.

■

Chapter 36

Quadrus

The Quadrus office complex is one of those places that is so quiet, you actually find yourself speaking softly so you don't disturb the peaceful tranquility. You half expect some New Age tingly music to be piped in, gently mixing with the breeze, and to find lots of cross-legged folks chanting "ommmmm, ommmmmmmm." It's kinda dull.

You could hardly guess that there is maybe $30 billion buried in the surrounding offices just itching to get out and fund something worthy. This is Sand Hill Road, the de facto capital of venture capital. Across the street, the Stanford Linear Accelerator is busy smashing atoms day and night, but here, there is that mellowness that drives entrepreneurs crazy.

So maybe this wasn't the best place to meet with Dr. Gary Glazer, but he insisted, maybe hoping that some of his excitement would rub off on someone around here, or at least his view on where medicine was going could gently mix in the breeze with all the money around here and something, anything might happen.

We lunched at the Quadrus Café, an overtly California cuisine place smack-dab in the middle of all these venture capitalists so they didn't have to start up their Porsches and risk door dings in the relatively untamed Palo Alto. Instead, they sip lattes and munch on bean sprouts and spout about their latest investments in interoperable enterprise software.

Dr. Glazer was a few minutes late and I ran into two different folks I knew, both of them walking double time, complaining about a board meeting running late and having to get back for a conference call. Now I remember why it's so quiet around here—nobody actually does much of anything but sit around in meetings.

"Sorry I'm running late." Dr. Glazer was walking toward me double time.

"No problem, I was just catching up with some old friends," I said.

"Some fires down at the lab—not real fires—just lots going on. Seems like a new breakthrough every day, it's hard to sort them out."

We sat down and I ordered a greasy cheeseburger and fries. It was my own experiment I was running with doctors—to see if they were true believers. If they thought heart and stroke and cancer were really curable, they would probably load up on buttery and grease-dripping food, maybe even take up smoking. What the heck—piss in the wind.

I checked out Dr. Glazer's face when I ordered: no grimace, deadpan—a good sign. But he ordered something with bean sprouts and lots of fruit. Damn. Maybe this stuff isn't just around the corner.

"Thanks for taking the time. I know how busy you are," I said.

"That's fine—I'm intrigued by your persistence, digging around in the future."

"I'm kinda stuck," I admitted.

"Welcome to the club."

"I think I can see where imaging is going. 64-slice, 256-slice. Maybe 1,024-slice. When it gets cheap enough, scanning should hopefully find stenosis and plaque and all the stuff in our arteries ahead of time, before it causes a heart attack."

"Or stroke. If you are right on the cost side of things, when it gets cheap enough, imaging will be widespread and find various problems—so heart attacks and strokes will become rarer!"

"Rarer as in uncommon, infrequent . . . obsolete?" I was going to work my way to this question, but Dr. Glazer brought it up first.

"Perhaps."

"Wow" was all I could say. He really did just say that two of the Big Three could be things of the past. Remarkable.

Gary hit me with: "But that's nothing." I wasn't ready for that line.

All I could picture was Dustin Hoffman saying to Robert De Niro in *Wag the Dog* as things got more incredible, "This is nothing." I'd just heard one of the smartest people in medicine say heart attacks could become as quaint as egg creams and Norman Rockwell, an earth-shattering perspective, but then add, "This is nothing!"

"Really? Imaging can do more?" I asked. I was prepared to talk about what 256-slice scanners could do and we were already way past that.

"Molecular imaging is the next step. That's what we are working on," Gary said.

"You mentioned something to me about this at that conference in San Francisco. I've been digging around but I still don't get it. You mean when you get to, I don't know, say 4K-slice scanners, you can view individual molecules?"

"No, it's not that." Ouch. Dr. Glazer had just deflated my balloon.

"It's not?" I probably sounded dejected.

"That wouldn't tell you much—it's the old needle-in-the-haystack problem."

"So imaging hits a wall?" I wondered.

"Let's look at it this way. A 10-millimeter tumor has around 1 billion cancer cells in it."

"Okay." If he says so.

"We can certainly find that with CT, if you look in enough places, but the tumor probably has already done damage—lots of symptoms, spread to too many places. A tumor the size of a pinhead has maybe 1 million cells. It's there, and not yet dangerous, but almost impossible to find with conventional imaging."

"But what is molecular imaging, then?" I still didn't quite understand.

"One important characteristic of biology is its diversity, its variation. It's why personalized medicine is so important. The key to imaging in this diverse world is to scan for structure *and* function." He stabbed at his salad as he emphasized each word.

"Function?" I asked.

"Today we're just imaging structure. Over the next five years or so, in humans anyway, we'll get to a temporal resolution of a couple of hundred microns. We'll be able to image any named vessel in the body. It will generate more info than the Human Genome Project."

"But—"

"But this is just gross organ function only. We'll see arteries to kidneys or the liver, but not what individual cells are doing. Sure, we can find soft plaque, that's easy."

At least he didn't say, "This is nothing!"

"Right now, we only really look for tumors in the colon and lung. You know why?"

I shook my head no. I felt like an unprepared med student.

"Because we can. They're sitting in an air pocket, so you can actually see them. You pump the colon with air before you go looking." I think I grimaced at the thought. "Tumors in soft tissue are tougher to find. There's just not enough contrast. Tumors in the liver, pancreas, ovaries, brain, good luck. It's just soft tissue on soft tissue. It all looks the same."

"Until it gets big enough?"

"It's got to be 10 to 15 millimeters wide before you can make it out in soft tissue. It's that billion-cell tumor, which, by the way, is already halfway through its life cycle. By the time you see it, it's probably already deadly."

"But scanning down to the molecular level—"

"No, no, we're working on a totally different approach. Let me put it simply. We can insert a probe. The probe finds the tumor when it is 10 million cells, 100 million cells, and then lights it up."

A probe like *Fantastic Voyage* with shrunken doctors looking for tumors? Luckily I was only thinking that, and didn't say it. But I must have looked confused.

"This is biochemistry, not some nano-robot sniffing around," Gary said.

Whew.

"Cancer is not one thing—there are lots of different types, each with its own unique characteristics. But one thing is the same—problems occur internal to the cell long before structural changes happen. Some abnormal functioning. Often, it is a protein that is overexpressed and messes with the normal functioning of the cell."

"Okay."

"And that's how we'll catch it. Early enough on, a blood test might pick it up. That's what CEA and PSA are for, markers for cancer."

"CEA?"

"Carcinoembryonic antigen, CEA. It's a protein found in embryonic tissues, and pretty much disappears by the time you are born. They found that many patients with advanced tumors have high levels of CEA. You can test for it in a blood test. It's pretty good at colorectal and GI cancers. But early tumors don't raise CEA levels, and not all cancers do, so it's not a great screen. Plus sometimes other things, even ulcers, can increase CEA. It's mostly used during therapy—if your CEA level drops, it probably means a tumor is shrinking or killed by chemo."

"And PSA is for prostate cancer?"

"Yeah, it's a marker—prostate-specific antigen. A high PSA doesn't mean you have prostate cancer, just suggests you are at risk. Sometimes it's just an inflammation or infection. These things aren't perfect."

"Is anything?" I asked.

"Nothing works 100%. You just try to get close before you start removing prostates, or someone's pancreas and put them on insulin for the rest of their lives."

Gee, I had just shipped a couple of vials of blood to have genetic tests look for mutated genes.

"You think genetic tests will be part of this?" I asked.

Dr. Glazer started to chuckle. "There's an adage in the research field that says that anyone who claims they have found a single genetic marker for disease gets four published papers out of it."

"Oh," I said. Maybe there was something to this.

"First, you publish a case report, then a report on a series of patients in a limited trial, then a review with a recapitulation of your claim, and finally"—he paused for effect—"a retraction."

Ouch.

"And molecular imaging? Probes? Will they be any different?" After studying the biotech mess and drug discovery and the disappointment with genetic engineering, I was becoming skeptical that anything thrown into your body is worth a damn.

"The proof will be in real humans instead of just genetic codes. With molecular imaging, you first do a blood test—

molecular diagnostic looking for any number of cancers. Maybe it's a few biomarkers, probably more like 20 of them. Put the thing on a chip and you can do 10,000." Did he just say chip? "That's how you get to the 90%, even 99% reliability."

"But that's just finding some expressed protein in the blood?" I asked.

"Right, that only tells you something might be in there. It could be anywhere. The next step is to custom-design a probe, with some radioactive or even fluorescent tag, and then inject it into the patient. You then do a PET scan or optical imaging and see what the probe lights up."

"These probes are . . . ?" I asked.

"Today, it's just FDG," Gary said.

"FDG?" Every time I start to understand how this all works—a curveball.

"Fluorodeoxyglucose. It's the only imaging biomarker approved. Most cancers require sugar, so FDG tends to find cancer and tumors. The whole PET imaging industry is driven by FDG. But it's not perfect, either. It tends to find only advanced tumors. Plus, your brain needs lots of glucose, too, so there is the risk of a coma with too much FDG."

"And there are better probes?" I asked.

"Sure, but not great, not yet anyway. That's what we're working on." Dr. Glazer sighed and then looked around the restaurant. "I wish there was a bigger effort. It's going to take a lot of capital to perfect these biomarker probes. But the market"—he paused and looked around again—"is going to be huge." I thought I saw one of the guys who was in a hurry earlier straining his neck to listen in, an important trait for venture dudes.

"You should meet Sam Gambhir. He's doing the most interesting work on molecular imaging. If you can get in his lab, you can see it in action."

"I'll call him," I said. I still didn't really understand what these probes could do.

"And when you can find these tiny tumors, then what?" I asked.

Dr. Glazer unleashed a huge smile as he pushed away from the table. "The trick is being early enough. Once you find it, you zap it, heat it, cut it, or freeze it."

He left with that explosive thought hanging in the quiet breeze.

Look for unique proteins, image for a tumor and then zap it out.

Screen, glean, clean.

I had some boning up to do on PET scans. I had heard the name but wasn't really sure what it meant.

As Groucho Marx used to say, sometimes a cigar is just a cigar. You can search away with CT and MRI through the body structures and find things you think are tumors, but are something else, a lesion, a tear, whatever—just cigars. But, worse than that, maybe something looks quite normal but is instead a deadly tumor. This is especially true in the soft tissue regions— even an MRI doesn't have high-enough contrast to find small tumors.

As cancer tissues divide and replicate rapidly, they are in constant need of energy, especially glucose. Introduce glucose into the body, somehow mark it with a radioactive substance and it will find its way to the cancer and light up like a spotlight telling you where the cancer resides.

That's the theory behind PET scans—positron emission tomography. It sounds easy, but of course there are lots of hassles. First off, introducing radiation into someone's body is not the nicest way to treat your patients. So you want to use some sort of radioactive gunk with a short half-life, meaning it stops being radioactive soon, hopefully by the end of the day.

You can't buy this stuff in a jar, you really have to make it yourself, so any PET scanning setup also has to own their very own cyclotron—that's right, a particle accelerator. Using strong magnets, you accelerate neutrons into a circular orbit, where they gain lots of energy and then, at close to the speed of light, get smashed into a target substance. Positively charged particles fly off—radioactive isotopes with short half-lives.

Fluorine-18 has a half-life of 110 minutes, carbon-11, 20 minutes and nitrogen-12 about 10. Using chemical synthesis, these isotopes are combined with some sort of marker that can be injected into patients. Don't try this at home.

The most commonly used marker is FDG, fluorodeoxyglucose, tagged with that fluorine-18 isotope.

Then you inject, wait 30 minutes or so, and put detectors around the body. When the positively charged positrons hit electrons in tissue, the are neutralized, releasing two gamma rays in opposite directions. This is what lights up your detectors or film—the intensity based on how many positron-emitting radioactive molecules end up at each tissue location.

"Dr. Gambhir, please." I lobbed in a call.

"Yes?" he answered.

"Oh, hi." I was expecting an assistant. "Dr. Glazer suggested I contact you. I've been doing some work on technology and health care and scaling and he walked me through some of the things you are doing with molecular imaging. I think I understand it, but can I come in and ask you a few questions? It seems quite fascinating."

"Sure, come on in. Have you seen any of the mouse examples?"

"The what?" I asked.

"We'll probably go through eight or ten thousand mice this year and do twice as many next year. We're only limited by the lack of conveyor belts to get the mice on and off the imagers."

Mice?

"Okay." I said tentatively.

"Before we meet, why don't you call Shay Keren. He works in the imaging lab. Have him walk you through it and show you what's going on. Then I think you'll have a better understanding of what we're doing. We can meet after that."

"Fair enough," I said. Seeing is believing, I suppose.

Chapter 37

■

Mousey-Mouse

The high-pitched screaming was coming from downstairs. I didn't know what the heck was going on. I was on vacation, just hoping for rest and relaxation, and this was definitely disturbing the peace. I bolted down the stairs and ran smack into my wife, Nancy, who happened to be the source of the screaming.

"Get it out. Go down there and get it out."

"Get what out?" I asked softly, trying the old calm husband thing. It didn't work.

"Get it out. There is a *mouse*. Ahhh." The word *mouse* got louder as she screamed it.

My boys were huddled around the door of the room, laughing. "Dad, there is a tiny mouse eating pretzels in my bed," my oldest son, Kyle, told me.

"Jeez." I didn't know what to do. My only experience with mice was watching Tom and Jerry cartoons. So I closed the door, hunted around for a broom, found it and prepared to attack.

"Go Dad."

I whacked the bed with the broom and the mouse leapt off into the corner. I threw the broom into the corner, but the mouse scrambled and darted into the bathroom. No way was I going to kill this thing.

"Okay, boys, we got Mousey-Mouse trapped. Close the bedroom door and open that door to the porch." They kept smiling because even with the door closed, you could still make out the screaming upstairs.

I swept the mouse out of the bathroom. It went flying and crashed about halfway up the wall, and then went back under the bed. I poked at it until it emerged along the outside wall. I stepped up and swept it toward the open porch door, which it hit, and then it turned and scrambled outside. My boys were applauding and I milked it for all it was worth.

I yelled out the door, "Get out. And you tell all your friends to stay away from here, too." Now my boys were on the floor, they were laughing so hard. A proud moment.

"And over here are our bioluminescence imagers. You can inject the firefly luciferase gene into tumors and then send in an enzyme that excites the gene and it literally lights up. This imager . . ." Shay Keren was giving me a tour of the Gambhir Lab.

I was barely listening. I couldn't take my eyes off a Plexiglas box labeled KNOCK-DOWN BOX. Inside were three pink-eyed white mice that had stopped scurrying around and were instead sprawled in the box, twitching away on their backs and sides.

I had gotten lots of tours of factories and semiconductor fabs and data centers. There were lots of blinking lights and whirring disk drives. Twitching mice were new on me.

"After injecting the substrate, you knock the mice down with isoflurane and place them in the Xenogen imager. We can do five at a time. You can see the little nose cones inside the box, you just lay the mice in there and keep pumping the isoflurane for as long as you need to keep them down."

I could see that Mousey-Mouse's friends had all been

rounded up and put in a basement lab in the Clark Building at Stanford.

I must have had a perplexed look on my face—twitching mice, firefly extract, what the hell were they doing in this place?

My tour guide, Shay, had a sly smile on his face. "Maybe I should show you how we used to image mice and then work our way back to these imagers. Anything you specifically want to see?"

"Okay. I've heard about molecular imaging, and was told this is the place to see it in action," I told him.

"We'll get there. Check this out first." We walked up to a waist-high white box, maybe ten feet wide and three feet deep, with a two-by-three-foot glass-covered opening. It was either an industrial freezer or a copying machine. I peered in the opening and saw a thin wire stretched across two metal struts and some other shiny stainless steel metal receptacles. This whole contraption looked like a giant high-end deli meat slicer.

"Is this—" I started to ask.

"Before we really perfected bioluminescence or any of the CT imagers, the only way to look inside of mice was to use this Leica machine. I think it's called a cryotomograph or something like that. You freeze the animal, kill it, slice it into 50-micron, maybe 20-micron-thin slices—"

"It slices, it dices—" I blurted out, but then my mind wandered back to the convicted killer article—the slices of Joseph Paul Jernigan. I felt my whole body tense up like it was frozen and then a shudder made its way through the system, snapping me back to reality.

"Actually, if you go too thin, it could take all day to slice up a mouse nose to tail. Okay, after you slice it, you place the slices on what's basically a wide piece of Scotch tape . . ." I looked into the cryo slicer and there were probably 20 pieces of tape, each with a nice mouse slice, sticking to the back side of the machine's wall.

"Okay, I get it."

"And then take it to a microscope or even a different imaging machine. This way skin and bones don't get in the way of the signal. We still use this every once in a while."

I had a new appreciation for Bracewell's algorithm and CT slices. Beat cutting 'em up.

"Let's go check out the other imagers." Shay headed toward the other side of the lab. I trailed behind him. One of the researchers in a white lab coat had a huge needle inserted in a mouse's tail and leaned toward me as I walked by and whispered, "Some of the folks around here refer to this place as Mouschwitz."

That's not even funny, but the name stuck in my head.

"Actually, we treat the mice pretty well here," Shay explained. "That isoflurane gas is pretty harmless. The mice live a lot longer and better with these new imaging systems; not many are destined to be sliced to pieces."

I didn't care either way. If my life was going to be saved someday because some rodent bit the dust, fine by me. Beats whacking them with a broom.

"Here's an atomograph from GE, probably half a mil. Think of it as a raster scan of the mouse. It's based on the time of flight of photons. Since light decays, you get data to calculate depth and can create 3-D models of organs and tumors."

We kept walking.

"Over here is a small CT machine. Pretty standard, low dosage of radiation, but there's also a therapeutic mode, a focused beam so we can zap stuff if we want to."

"Looks like you could fit a dog in there," I said.

"Pretty much any small animal is fair game," Shay replied. I was learning to appreciate his detachment. We walked through a door with a radiation sign taped to it. I thought I saw a piece of mouse tail hanging from the tape. I don't know, maybe I was seeing things.

We walked past little white cutting boards with wires attached to them. Shay saw me looking. "Those are warming beds—for surgery. Now, you know about PET?"

"A little bit." I said meekly. I was hoping no one was going to do a mousey-rechtomy on one of those warming beds any time soon. Or cut open a Boston terrier. Around the corner was a two-foot-diameter fortress, made out of what looked like lead bricks.

"For kryptonite?" I asked, pointing.

"I'm not familiar with that compound. PET works by reading the gamma rays from a radioactively tagged marker. We use FDG, basically glucose tagged with F-18. Then we image in here." Shay pointed to a loop of metal with wires heading to a white box sitting next to it. "It's mainly for rodents, but we've been known to scan a squirrel or two. We can just barely fit a monkey head into this thing. You've got to, uh, monkey with it a bit, but it works. You inject the marker into the tail or vein, knock 'em down, and wait 30 minutes or so, the marker then goes to where it's supposed to and we can image for it.

"This kind of imaging shows metabolism. It's functional, not structural. It's the first step towards molecular imaging."

"Okay." Now we were getting somewhere.

"We don't like to brag, but this is probably the biggest small animal imaging lab anywhere."

My head was nodding. Shay had that smirk again—kind of a "welcome to my world" look. "Any questions?"

"I've got millions of them," I said.

"Why don't you just sit here and watch a few of our researchers do some imaging. Maybe it will make a little more sense."

Not much was going on right then, so I looked around to take it all in. Next to me was what looked like one of those brown refrigerators you see in dorm rooms. On it was taped—

no tails—a sign that read NO DEAD ANIMAL STORAGE IN THIS FREEZER. NO FOOD OR DRINK IN THIS FREEZER. That goes without saying.

A young woman walked in with a Bloomingdale's brown shopping bag. She began pulling out what I at first thought was her lunch, but instead were several Tupperware-like containers each containing five mice, food pellets, water and I guess kitty litter lining the bottom.

Shay leaned over. "We've got six holding rooms just over there, probably several hundred animals. It's connected via tunnels to the main animal storage facility on campus. You don't even want to know how many there are."

No, I didn't.

The researcher pulled the mice out one at a time, popped them in the knock-out box, and turned on the isoflurane. After what seemed like forever of twitching, she injected something in their tails, marked the tails with an indelible ink Sharpie—didn't see if she pulled it out of her sock—and then placed each mouse in a nose cone in the Xenogen box. Then she started clicking around on the attached PC, setting the image time to five minutes, and we waited. Well, I waited. The researcher got a paper towel and cleaned the, uh, Raisinets, from the bottom of the knock-down box.

Five minutes later a red, yellow, green and bluish image showed up clustered around the belly of each mouse.

"Those mice are genetically engineered to have the firefly gene. The injected material is meant to show up in some specific spot, looks like the liver in these. It carries the enzyme that turns on the firefly illuminant. So this way we can see if the injected material finds the liver or goes to other places as well. The trick is to get it to localize."

Another woman with a white lab coat entered the lab, pushing a cart. It looked like a dim sum cart to me, but the dishes stacked up were more plastic mice pounds.

But these mice were different. They were pink and completely hairless, victims of a nuclear accident or something.

"Nude mice," Shay whispered.

"I can see that," I said.

"No, no. These are mice specially bred to have no immune system. This way they don't reject human tumors or disease that we might inject into them. These probably have cells from some cancerous tumor injected into a kidney or liver."

"Then what?" I asked.

"Well, that's the reason to have all this expensive stuff. She is probably injecting some compound to see if it finds the tumor. The image will light up around the organ if it does. Drug companies send us drugs sometimes to test their effectiveness in finding specific tumors. It beats spending hundreds of millions on clinical trials without really knowing. But this one looks like an MI experiment."

MI. I assume he means molecular imaging. I still wasn't sure what that really meant.

But seeing all those syringes around, Hmmm. I might need more blood drawn.

"So, you guys can draw blood?" I asked. I remembered how hard it was to get my blood drawn and was jealously eyeing the syringes.

"From mice?"

"Well, from hu—from me—uh, let me retract that question." I scrambled for the exit.

On the door on the way out was a sign mimicking those milk ads that said GOT MICE? PLEASE CHECK THAT YOU HAVE ALL YOUR FRIENDS BEFORE LEAVING.

I chuckled, but as I left, I thought I heard one of the mice in a Plexiglas box say, "Get out. And tell all your friends to stay away from here, too."

Part V

Chapter 38

■

Genetic
Test Results In

I just had to ask, didn't I?

It took weeks and weeks, but my test results finally came in. I felt like a senior citizen, hovering around the mailbox, waiting for the mailman to show up, asking Nancy if the mail had arrived.

The first results were the genetic test for cystic fibrosis. I ripped open the envelope and started scanning as fast as I could.

1 in 3,000 Caucasians contract the disease . . . Check.

Damage to the respiratory system . . . I'm always bent over playing hoops. Check.

Inherited disease of the mucus glands . . . abnormally thick, sticky mucus . . . See hoops above. Check.

Problems with digestion can lead to diarrhea, malnutrition, and weight loss. Are cheese steaks malnutrition?

Inability to conceive a child . . . I've got four boys. Hmmm.

Our identification of the cystic fibrosis transmembrane conductance regulator, ATP-binding cassette (sub-family C, member 7) and conditions of mutations in the CFTR gene . . . Mutation? Gulp.

Your prognosis: Negative.
That's good, right?

I'm feeling a little better, anyway. I'm not going to be hocking up huge spitballs and my kids aren't genetically susceptible to cystic fibrosis. I wonder how many more of these diseases I can tick off before I stop worrying about losing the genetic game.

Fortunately, the next test results to arrive had Veterinary Services in the return address. This ought to be interesting.

"Your Boston terrier," it said, and then an underlined fill-in-the-blank spot had the name Fluffy written in ink, "is not genetically inclined to carry Type I von Willebrand's disease." The rest of the letter rambled on about the accuracy, or lack thereof, of genetic testing and gene mutations.

And then came the shock.

At the bottom of the letter was a handwritten note: "I'm sorry to inform you that your dog has tennis elbow."

Well, at least someone had a sense of humor.

A couple of days later, I received a letter addressed to Dr. A. Weisenheimer.

Inherited gene mutations have been shown to cause hereditary nonpolyposis colorectal cancer. Mutations in the genes for HNPCC increase an individual's risk for colorectal cancer by 70% to 80%. Further, the risk of colorectal cancer and endometrial cancer before age 50 is increased to 20% to 25% compared to just 0.2% in the general population.

This is pretty serious stuff. One mutated gene and my risk goes from 1 in 500 to a 1 in 4 chance of getting colon cancer. Is that right?

The blood sample of your patient, *George O. Jung-el,*

I forgot I had used that name.

has tested POSITIVE for inactivity of the MSH2 gene.

What? Positive? That's my blood.

MLH1 and MSH2 are known as "DNA mismatch repair genes" for the protein products of these genes and their responsibility in repairing mutations that occur in the genome. Inactivity is associated with mutations in other genes, a "mutator phenotype."

So I have a gene mutation that means other gene mutations don't get repaired. What kind of circular nonsense is that?

Is this a problem? Do I have colorectal cancer? It doesn't say that. It just says my risks have gone up. What do I do? Should I become a vegan? I think I'd rather staple my eyelids to my forehead than become a vegan.

But then again, it's not like I can go have my colon removed. Or just part of it? A semi-colon? Isn't there a writers' joke about removing the appendix? This is no joke.

Or is it? Maybe the test was wrong. It was such a hassle to draw blood—should I do it again?

Should I call my parents and yell at them? That would be a first. I'd feel better, but that's about it.

Not knowing what to do, I did some more research. A study of Finnish patients showed these mutations were responsible for 3.4% of colorectal cancers in that country. Maybe it's from reindeer meat. Who knows?

I had heard more than once of women who had cancer in their family, took genetic tests for breast and ovarian cancer, and when the results came back positive, quickly went out and got both a double mastectomy and a hysterectomy. All based on the increased risk and probability of getting cancer. It sounds extreme, but I'm not judging—I might have done the same thing.

My mind kept racing. Could I know right now? Or was this a ticking time bomb? I supposed I could go back to my doctor and ask, but he might get out the rubber hammer again and whack my knee, $400 dollars, please.

Colonoscopy? Probably. I had to do it anyway at 50. But when digging around about CT scans, I read an article that said that 23% of colon polyps are on the outside of the colon. Plus, there can be complications. Of course, this study was funded by a group pushing virtual colonoscopies, but it made sense, why should cancer be just on the inside of the colon, it's just tissue.

I could only imagine the dreams I was going to have tonight. I suppose Jack Daniel's might keep the scary ones away for a few nights while I did some more digging around.

Every time I searched for colon or colonoscopy, I kept seeing ads for stool occult blood tests. And it wasn't Halloween. I finally figured out that *occult* means "hidden." One health supply website says:

> Convenient, Sanitary—Easy throw-in-the-bowl test. No messy handling of stool. Pad will turn blue/green if blood is detected.

No messy handling of stool—kind of a personal motto of mine. $12.74—hey, that's 30% off. I clicked the "add to cart" button and noticed that shipping and handling was almost as much. What else they got on this site? Hmm. Cholesterol kit . . . click . . . $14.99—that's 40% off, man, am I saving money! Personal hormone testing kit, $6.47. I'll take that, too.

I skipped the antioxidants, the male enhancers and the

weight-loss products, but was intrigued by the section on magnetic wraps. Two in particular looked interesting. One was a magnet headband, $7.16. For no particular reason, I chose red. Two removable 1-by-1-inch strips of five magnets—800 gauss and all bionorth (–) facing the body. Cool. Whatever the hell that means.

> This item may be helpful for muscular tension headaches. Experimental device: more research is needed, results vary from individual to individual. This product is not intended to diagnose, treat, cure, or prevent any disease.

For $7.16, I was willing to take a chance. Another magnetic wrap made me laugh, and certainly took my mind off the fact that I may have a genetic mutation and I may have some colon cancer mutating while I waste my life surfing the Web.

The hell with that—I started to laugh louder.

Sale price: $8.76. A magnetic tennis elbow band.

Too perfect. I bought a matching set!

Chapter 39

■

Med Conference

I was back in New York again. The American Airlines shuttle across the U.S.—another three mrems of radiation.

The Grand Hyatt hadn't changed much. Same crowded sidewalk next to Grand Central Terminal. Same bustling lobby. Same set of four or five escalators up and over and up and over until you reach the ballroom level. As with every investor conference, the ballroom was filled with scurrying mice in tailored suits, all looking for the next greatest thing. And for days, every 30 minutes in six different tracks, companies endlessly pitched their golden prospects. Every damn one of them sounded like the next greatest thing, so the mice scurried some more, discussed among themselves and probed and prodded CEOs with insightful questions until the mice were so confused, they just asked someone else what they should buy.

I'd been to so many tech conferences I could tell you where the bathroom and best cellular reception is in the leading hotels of the world. But this was a health care conference. I'd pulled some strings at Bear Stearns and wangled an invite. I figured it would be a great forum to get instantly up to speed on the health care industry. So I put on a suit, slogged to New York, sweated in 95-degree heat, and pretended to be interested in Phase II trials of monoclonal antibodies. What I learned was that health care investing was a bigger crapshoot than technology.

"What do you think of these guys?"

"Curagen?"

"Yeah."

"Too early."

"But they're in Phase II on their cancer thing—PXD101, I think. I thought you liked Phase II."

"I do, but it's not theirs."

"Sure it is."

"Nope, they licensed it, not enough upside. No one is going to care."

I was eavesdropping on two guys I recognized from the past. One guy used to run some tech money, I thought I remembered him getting blown out in 2000 for being long and wrong. But he was obviously still in the business. I strained to read his name tag, without luck. The other guy was a classic popcorn hedgie. Always showing up where the action was, figuring out some angle and trading rapid-fire looking for returns.

"For me, it's just one thing."

"Market size?"

"Nope, these are all billion-dollar markets."

"Management?"

"Nope, they're all trying to hit fish with a baseball bat."

"True. So what is it? Balance sheet? Partners? Their logo?"

"You just play the trials. I don't give a shit what drug trial it is, it almost always works the same way."

"Get out of here."

"No, check it out. Preclinical is bullshit, you could get gum off the sidewalk to reduce tumor size in rats. So what? Even Phase I is a load. They are mostly bent around to get decent enough results so you can get to Phase II. I've tried to time Phase I announcements, but no dice. You always wait until you are about a third of the way into Phase II, then you buy the shit out of the stock."

"C'mon."

"It works. The first few press releases are almost guaranteed to be positive. Companies will do anything to get their stock up before the final results hit. Phase III is just so damn expensive, they want to sell stock or converts or something so they don't have to give the damn drug away to Big Pharma."

"Okay."

"Look, the stock almost always pops. I usually sell into it, unless the press release is blow-away efficacy numbers; then you can stick around. But not forever."

"No?"

"Hell no, you almost always sell right before the Phase II trial is over. Everyone figures it's gonna work, so the stock goes up until the day of the announcement. It almost always gets whacked when they announce the real results, people already know. The doctors have been trading the stock ahead of you and dump it on the news."

"That's it."

"I sometimes do Phase III, but that's harder. Even if something works, the costs kill you. The stock's going down."

"But it pops on the news at the end of Phase III."

"Sometimes. Not always. You can get killed if the trial fails. Maybe 20 drugs will be approved this year."

"Can't you play that?"

"Nah. I don't like to gamble. Not even in Vegas."

I spent several days walking the halls, popping into presentations, seeing if I could hear about some sea change in health care. Investors always hear about things first. Instead I heard about drugs that might or might not treat bone metastases, myelogenous leukemia, multiple myeloma, kidney inflammation, and on and on.

I didn't hear about innovation—more like regimentation.

I don't know how many times I heard the expression fully randomized, double-blind, placebo trial. Every biotech and pharma company pitched some set of compounds in trials. They were rooting away, but had absolutely no idea if the damn thing really worked. Even big companies with billions in sales of some drug pitched their replacements when they went off patent. It seemed like a giant roulette table. If 17 black comes up, you win 35 to 1, otherwise, zippo.

Even the popcorn hedgie didn't like those odds, settling for smaller returns with a bet based on knowing how fast the wheel spins. Jeez.

One of the things I kept hearing about in presentations was Pfizer and some RNA thing and some new drug they had in the pipeline. Presenters would say, "We are working on RNA interference, not much different from what Pfizer is doing."

Again and again. Between meetings, I heard a couple of folks throwing around big numbers and Phase III trials and cholesterol, so I started eavesdropping again. Nothing is secret at these conferences!

"They're talking about over a billion dollars just for the Phase III trial. I'm not sure anyone has spent that much," one guy with a goatee was saying.

"I thought Lipitor had a few years to go on patent," said another with a soul patch.

"It does," goatee explained, "but Zocor's off in 2006. I think Pravachol is, too. Those two going generic means seniors are going to switch. Wouldn't you pay 35 cents versus a couple of bucks for Lipitor?"

"Sure," said soul patch.

"That's why Pfizer's dropping the billion on Phase III. I'm not even sure this RNA interference thing really works."

"Why not?" asked soul.

"This whole thing of turning genes on and off makes sense. You muck with the RNA before it turns into proteins. In this case, it's the CETP gene, cholesteryl ester transfer protein, I think," answered goatee.

"You just shut if off?"

"Pretty much. When you do, your HDL, good cholesterol, goes up. Pfizer has this drug, torcetrapib, that seems to work. That's why they're going to drop a billion plus on Phase III. They should know in 2006. And then they bought Esperion, which has some synthetic HDL they make out of peptides."

"They've got their bases covered," said soul.

"Even better. You need a statin to lower your bad LDL and this new torcetrapib thing or the peptides to increase your HDL. Bingo," goatee practically shouted. "Match, set, point."

Wow, the cholesterol conspiracy lives on. This drug sounds pretty neat. I just wonder if it reduces heart attacks. At the end of the day, that's all that really matters.

There were the two hedgies again.

"Stem cells," one of them said.

"What about 'em?"

"You figure out how to invest yet?"

"Lobbyists."

"Thought so."

"Look, they may work someday. Rebuild organs. Get the crippled to walk again. Solve Alzheimer's."

"Yeah?"

"But if there was a public company today, I'd short it. There's plenty of time to see if they really work or not. I suspect we'll see $10 billion thrown down the drain on false hopes—California is $3 billion alone—before we see real companies figure out how to harness it. Stay on the sidelines."

•   •   •

I still had the rest of the afternoon to kill and I was getting tired of Phase II and Phase III clinical slides. But they were everywhere. I started picking which company to see next based on how cool their name was. Downstairs, tucked into a tiny conference room, was Stereotaxis. Sounded good to me.

I walked in late and made a commotion tripping over just about everybody on my way to the only open seat in the front row. This meant I had to pay attention.

"And then once we are there, we can map the entire region in a second or two, creating a 3-D map of the areas that require arrhythmia ablation. Then the doctor can guide the tip remotely to these regions and apply the electrical pulse that destroys just the heart muscles that are responsible for the irregular heartbeat."

"Can you describe the navigation system one more time?" someone asked from the back of the room.

"Sure. We move in two magnets, one either side of the patient. We can work with all the major fluoroscopic imaging systems, without getting in the way. What we do is put a magnet in the tip of the catheter. A doctor in an adjacent room, behind lead glass, using just a joystick, can maneuver the catheter through the arteries. It's a bit like a video game."

3-D mapping. Video games. I'd seen it for imaging, but not for moving something around inside.

"You may not know that some percentage of coronary catheter procedures end up with a perforation. Maybe up to 1% of them."

This I didn't know.

"We strip out all the mechanical attachments from our catheter. We're left with just a hollow tube with a $10 micro magnet on its tip. The thing ends up like a wet noodle—much more flexible and less likely to perforate anything. And you don't need three doctors to guide it, just one guy with a joy stick. We can even make 90-degree turns with our catheters. It takes the guesswork out of the cath lab."

I remembered seeing three doctors moving guidewires in and out of Mrs. Nussbaum in the cath lab by, what, guess-work?

"It's a big market. There are something like 6,000 cath labs, over half of them in the U.S. They have an eight-year re-placement cycle, plus there are more and more cath labs put in, so we hope to be in close to 900 new and replacement labs."

They could probably pay for themselves just by getting rid of one of the doctors in the lab.

"There are 2.6 million procedures each year, some 15% to 20% are considered tortuous navigation. That's our primary market, but we think we can change the entire interventional cardiology market, with both easy navigation and an interven-tional suite for 3-D mapping for ablation and other things. I can't stress enough the value of real-time 3-D mapping of the heart and ventricles and then making decisions right then and there on strategies for intervention."

"Anyone else doing this?"

"There's Hansen Medical, from the folks who founded In-tuitive Surgical. But we think we have a lead in getting approval for the U.S. market. We have already changed the game in Eu-rope, especially Italy, and even Canada."

A bunch of magnets both augmenting and replacing doc-tors (and cure tennis elbow!). This is starting to catch on.

Chapter 40

■

Cancer Panel

Half of you in this room will get cancer. A third of you will die from it."

That's a pleasant way to start a talk. The speaker was Dr. Larry Norton from Memorial Sloan-Kettering. I suppose he should know. I looked around, hoping it was the other side of the room that was doomed. Then again, I was the one with a gene mutation.

He went on to complain about the drug discovery funnel. Tens of thousands of compounds go in and just a small handful come out. Biological concepts turn into lead compounds and formulation and safety testing and patent filing and clinical trials covering 2% to 3% of candidate patients and years and billions of dollars—it's a miracle anything comes out intact. His complaint was focused on Phase IV trials, combined treatments, cocktails, both the hassles in testing and the intellectual property monster that's created in the small event something actually does work. Dr. Norton must have used the words "restrictive environment" a dozen times.

"Oh, yeah, and $82 billion worth of drugs that are currently being sold, that successfully made it through this funnel already, are going off patent."

Ouch. No wonder there is such a huge scramble to get

through Phase II and Phase III and get a new set of drugs back "on patent" before the generics kill the entire industry's profits.

Dr. Norton went on and on and I started to tune out. The room was dark, it was late in the afternoon, and I'd heard these complaints before. Like the weather, no one does anything about it. Hurricanes still happen.

It's almost as if rational drug design meets the Heisenberg uncertainty principle. You can't know with full precision the values of certain pairs of observable variables. To paraphrase old Werner, the more precisely the gene is known, the less precisely its effect will be known in this instant. And you know those Germans like to be precise.

Will it change? Probably not. Just as the industry—meaning doctors, pharma companies, researchers, everyone—is whipped into a crescendo of whining, a Vioxx comes along and shuts them all up and the game is reset.

Vioxx was one of a new breed of drugs—COX-2 inhibitors. Some forms of inflammation can be traced to the expression of the cyclooxygenase gene and COX enzymes. Cut back on these enzymes, and voilà, pain goes away. One of the effects of simple aspirin or Advil is to inhibit these COX enzymes. But there are two types: COX-2, which is involved in inflammation, and also COX-1, which when reduced causes your stomach to bleed like a punctured tomato, which is why there is a limit to how many aspirin you can take.

Vioxx, Celebrex and Bextra were created to inhibit only COX-2—keeping the COX-1 around to limit gastric bleeding. A wonder drug, especially for arthritis patients who could take only so many Bufferins. Well, Vioxx was a wonder drug until it got overprescribed and some patients started having heart attacks and strokes.

Hmm, the human body really is a complex system. Push down somewhere and something else pops up to bite you. Merck withdrew the drug and Wall Street whacked $25 billion out of the value of the company. This was a surgical strike by investors figuring the value of the soon-to-be-missing Vioxx profits plus the potential payments in the lawsuits from class action cobras like Mark Lanier that were certain to follow. All in the first five minutes of trading. Quick analysis.

Dr. Gina D'Amato from the H. Lee Moffitt Cancer Center spoke next and spent 20 minutes talking about all the potential ways to fight cancer. My pen ran out of ink, I was writing so fast.

Signaling inhibition was big. P13K/Akt pathways were active in 70% of all cancers. Cut off the pathway and you cut off cell growth. Maybe.

Drugs that inhibit mammalian target of rapamycin, or mTOR pathways, are promising. Side effects? We'll see.

Toll-like receptors are an old method coming back in style. Fight cancer by injecting bacteria or some other virus that excites the existing immune system to do something about it. Toll is from the German word for amazing, but if it works, certainly someone will be collecting one. Too early to really toll.

Human Genome Science, one of the wonder companies that had hoped to leverage the sequencing of the human genome, had just blown up. They had found a target protein called beta-lymphocyte stimulator, BLyS. It keeps a certain white blood cell alive that when overexpressed causes the disease lupus. So HGS created an antibody to neutralize BLyS. It failed in clinical trails to reduce lupus symptoms. Back to the drawing board.

Viagra can lead to blindness. Those in the Borscht Belt might say, "Can't I just use it until I need glasses?" Type 2 diabetes drug Pargluva may increase cardiovascular risks.

Push down, pop up. You never know.

Meanwhile, Dr. D'Amato was frustrated. "Every time these potential treatments fail, I have patients dying for lack of better options."

•   •   •

Ribosome synthesis, angiogenesis, RNA interference, there's no end to the possibilities. Some will certainly work, but it will take time and energy and some success and probably lots of failure. That funnel gets very, very narrow.

Billions are spent, yet a third of us in the room may still die from cancer.

I may have some stupid gene mutation. Can't some drug un-mutate it?

There are hundreds of thousands of smart people working toward this goal, but it's a moon shot. My grandmother used to wonder if a bunch of hurricanes that hit in the 70s happened because astronauts "touched something" on the moon. It was hard to argue with her. "I not only read it somewhere, I heard it somewhere," she told me. The complexity of our skin and bones and blood and guts seems to fit my grandmother's concern. You "touch something" by mucking with cell growth and pathways and gene interference and you might get hurricanes elsewhere in the system.

Of course, my grandmother was wrong.

So is there a silver bullet, that magic cure, a red pill, or is it a blue pill, that changes everything?

Man, I hope so. But it ain't so obvious sitting here in the dark. I'm losing hope.

Chapter 41

■

Listwin at Buck's

I never turn down a chance for breakfast at Buck's in Woodside. Any place with a motto of "Flapjacks and Tomfoolery" gets my devotion. It's a daily ritual for venture capitalists—the khaki patrol is out early, snagging the tables and booths with the best views of the room so they can spy on who is talking to whom and have something to gossip about with their venture partners over lunch. I come here every so often, but it seems that the same folks—hey, there's Steve Jurvetson doing another nanotech deal—are here every time I come, almost no exceptions. Over the years, I've gotten to know Jamis, the owner, proprietor, and self-described Pancake Guy, and he made sure I had a table with a view of the viewers.

"Hey, sorry to be late. Don Listwin," he said as we shook hands and he slipped into the booth. I could almost hear the gears sputtering and subtle whispers, "Hmm, hedge fund guy and networking exec"—avert gaze—"wireless, maybe, how 90s, nothing of note here. Better to check out that other table with blog index guy and venture stud dripping yolk on each other."

"No problem," I said, coming out of my Buck's hallucination.

"You're a tech guy?" Don asked.

"Yeah, double E, worked on Wall Street for a bit too long."

"I'm an electrical engineer, too," he told me.

"I figured that. How'd you get into funding medical research?"

"I've had a lot of cancer in my life." He sighed. "My dad survived colon cancer. My mom died from ovarian cancer."

"I'm sorry."

"I don't know how much you know."

"Not that much," I admitted.

"Ovarian cancer is known as the silent killer. It's asymptomatic. Most doctors diagnose a bladder infection and prescribe antibiotics. It spreads fast. You can't even see it on scanners—it's soft tissue on soft tissue, kind of blends in. I figured there had to be some way to detect it early.

"Anyway, I left Cisco in 2000, had to sell the options at, well, you remember what it was like then. But I set up a foundation and figured I could take an active role."

"But how? Isn't most research segmented into silos, islands, funnels, whatever?"

"Look, you've been in Silicon Valley for a while, right? Stuff happens, technology spreads, people share information, you motivate people and they break down old ways of doing things."

"Sure." And how.

"Not much of that in medicine. The FDA is like the FCC, Big Pharma like the regional Bells."

"I get it," I said. His Cisco was showing.

Our food came. I had ordered the breakfast burrito, which is a grueling way to start the day. Don passed the food order test, eggs and bacon, excellent, a true believer!

"We're just missing entrepreneurs and open standards to get things going. I started Canary Fund to see if I could create an open platform for early detection. I work with Stanford and USF and the Hutch. It's just not there yet."

"Tell me about this Canary Fund. It's a venture fund? And what is the Hutch?"

"No, no. Let's see if I can explain. In the fall of 2003, I gave $10 million through a family trust to the Fred Hutchinson Cancer Research Center up in Seattle, to focus on early detection. I also joined the board to make sure it happened. They're the largest U.S.-funded cancer center, something like $300 million."

"Fred?" I asked.

"Fred Hutchinson was a baseball player who died young and his brother was a doctor who helped set the place up in his name."

"I had no idea," I said.

"They're not as famous as MD Anderson or Sloan-Kettering, but they do some important stuff. It's just that none of these places spends much on early detection."

"Why not?"

"Lots of reasons—politics, too hard, not glamorous, who knows."

"Okay," I said.

"You should meet Lee Hartwell. He won the Nobel Prize in 2001 for figuring out how cancer develops in yeast. He's attracted a lot of smart people there. He gets early detection."

"I still don't get how it works," I nudged.

"The keys are these biomarkers. You know what they are?"

I shook my head no.

"Something that you can identify in your blood or urine that points to some type of cancer being present. Oversimplifying, each tumor will express a unique protein, and if you can test for that protein, you may have a biomarker."

"Sounds straightforward."

"Not even close. There even is a biomarker for ovarian cancer called CA-125. It's 80% effective, meaning 20% false negatives. Let me say that again—it means that 20% of women who actually have ovarian cancer are told they don't with CA-125."

"Isn't that the same as PSA for prostate?" I remembered the 80% number.

"Yeah, pretty much. None of them are all that good. That's why I am funding this work—not only to find biomarkers but to create a platform so that others can more easily identify new biomarkers. Then medicine gets turned completely around."

"I still don't see how," I said.

"Well, you have to find the proteins. But everyone does different work on different machines, with results that aren't repeatable. Not great for bringing a true cancer screen to market. So we try to standardize simple things, like calibrating machines, like creating a real database of proteins and on and on.

"To get biomarkers to be real, you have to have both specificity and sensitivity. Picking up on just one protein and not missing it. But there are millions of proteins—we probably know about 5,000 of them. I'm funding a bioinformatics platform to hold all this protein info. It's amazing no one has done this yet. I've got a bunch of ex-Microsofters writing code."

"Why care about all those proteins? Don't only a few work?" I asked.

"Sure, but which ones? Turns out that if we use two markers—CA-125 and something else—maybe we can get the effectiveness to 0.95. With a panel of 20 biomarkers—maybe 0.99, 0.999. That's a huge change."

"Not easy?" It was a stupid question. Of course it's not easy or we wouldn't be sitting here.

"You are probably used to thinking of eight-inch wafers and billion fabs cranking out microprocessors. Forget all that. Think mass spectrometers, one protein at a time, often a protein whose quantity is eight orders of magnitude lower than other stuff in your blood. It's not easy at all. It's a couple of gigabytes of data out of the mass spec for each one."

"Does imaging fit in with all this?" Proteins didn't sound like chips that get cheaper.

"Sure, that's the end game. If your biomarker screen is positive, you can tag the marker and image for the tumor. It's what Sam Gambhir is doing over at Stanford."

"I've seen some of his stuff." Twitching mice, actually.

"It's hard to explain over breakfast. You should head up to the Hutch and check it out. It's not like anything else. It's more high tech than you might think."

"Really?" I didn't see how. "I may take you up on that."

"Sam is talking about getting it down to a chip—that's when it gets interesting."

Don Listwin paused and quietly looked around the room at the venture capitalists who were still stealing glances at our table, trying to read lips, perhaps.

A chip? What did he mean? Was there some chip that could do this stuff? Or just a piece of glass like the DNA stuff? I was about to ask when Don sat up and collected his cell phone and got ready to go.

"If I do this right, it then becomes investable. These venture guys can't see a return right now. I have to create something so that a huge return will be obvious; then the capital will come. They'll do it—you know how it works. But I figure someone like me has to create the platform. Once you show it works, early detection will be a huge business. After that, money won't be a problem. These guys will come out of the woodwork."

•   •   •

Will they? It's not like more money is going to pour into health care. At 15% of GDP, there's not much room to grow.

The same was said about personal computers. In the early 1980s, gobs of money was spent in corporations on information technology—mostly on IBM mainframes. For almost a decade, overall spending went sideways. But if you looked inside, the dollars were starting to be spent elsewhere—on the edge instead of in the middle. On personal versus enterprise-wide computing. It was a great time to be investing in software and chip companies—anything that touched personal computing and soon network computing.

But what about medicine? No one really wants to invest in doctors, but they do want to invest in intellectual property. Drugs, medical devices, that kind of stuff. But add another twist. Wall Street hates price controls and never pays up.

But what if the spending was on detection instead of intervention? With some breakthrough, the economic consequences can be staggering. If medicine as we know it is replaced by health monitoring, hmmm . . .

Chapter 42

■

Lighting Up

I walked through the ground floor of the James H. Clark Building. I was still annoyed at Jim Clark for being thrown out of his deal—but the building was real nice. The sign on the door said GAMBHIR LAB, and as I wound my way through, I saw a lab stocked with beakers and test tubes and a bunch of high-end computer displays. I guess that's science these days. I kept winding my way though, until I got to Sam's glassed-in office tucked in the back corner. It was what I would imagine any scientist's office looks like, stacks of magazines, books, a whiteboard and a huge computer display.

"Thanks for the mouse education," I said to Dr. Gambhir.

"I thought that would be a good place to start. There are a lot of issues, but we're making lots of progress," Sam said to me.

Sam was younger than I had thought, with glasses and a serious look on his face. I sort of expected frizzy hair and a wide-eyed, unkempt look and wrinkled clothes from the 70s. Nope, Sam was modern science—and from what I understood, a combined M.D.-Ph.D.

"I apologize that I don't have a lot of time. Perhaps we can talk for a while and then I can get you to come back when things are a little less hectic. We've got a big grant we're about to file. Lots of work."

I wondered if all these guys did was submit grants.

"That's fine. For now, I really just have one question. I was watching the mouse imaging downstairs last week. Bioluminescence, I think it was called?"

"Yes," he said.

"And I was told that they were transgenic mice?"

"Yes?"

"They're crossed with fireflies?" I asked.

"Sort of," Sam said.

"And then you inject them with an enzyme and only certain things light up?"

"That's right. The firefly gene is everywhere, and the enzyme goes anywhere, but we can turn on the promoter gene for just the things that we want to see, like cancer cells in the liver or kidneys or wherever," Sam answered.

"Okay, I sort of understand that. But when Dr. Glazer described molecular imaging, he talked about probes injected into humans that would light up cancerous tumors," I said.

"Of course, so what's your question?" Sam got to the point.

"Well, I remember the movie *The Fly*. Vincent Price and all. Didn't turn out that well. Do humans have to be genetically crossed with fireflies to get this stuff to work?"

Sam Gambhir started laughing.

"No, no. We use bioluminescence with mice because it's cheap and easy to do. Did you see the MicroPET machine in the back room?" Sam asked.

"I think so." I sort of remember—an opening the size of a monkey's head.

"Did you see it in action?" Sam asked.

"I don't think so, just the weird fire-mice-flies," I said.

"Okay, you've got to go back into the lab. Let me make a call."

He clicked on his PC and I saw the name Xiaoyuan Chen come up. He dialed the phone and left a message asking him to meet us at his office and show me something about copper-64.

"I think we're doing RGD peptides imaging today. Then you'll see molecular imaging in action," Sam explained.

While we waited, Sam started talking about modalities and probes and fluorine. I followed as best I could.

"These probes are the real key to molecular imaging. I understand Dr. Glazer walked you though this."

"Some of it. I'm still a little unclear on . . ."

"We can do a test ahead of time to determine which protein is expressed and then design a custom probe to seek out those cells. The test might be an ELISA but someday will be a chip that tells us exactly what to look for."

"A chip?" I've heard about this. Maybe Sam could tell me more.

"I can't talk about it that much, not yet anyway. It's pretty exciting. That's the grant I was telling you about—binding antibodies, quite exciting."

I was even more confused. Best go back to imaging, which I had some tiny grasp on. "So the probes are key?"

"Oh, yes. Really, we can custom-design probes and tag them with very small traces of radioactive materials, which the PET scan can pick up. Hopefully you'll see that in mice. With humans, there are regulatory issues, you know. The FDA considers anything you inject in humans as a drug and there is a long and costly approval process and—"

"The funnel," I said.

"We think we have a way around it, but it really shouldn't be an issue. These probes are not toxic at all, and the radiation is less than a third of a bone X ray. That's nothing."

The phone rang.

"Okay, let me have someone escort you downstairs to the imaging lab."

Damn, more twitching mice. I wasn't quite prepared for this. A 30-year-old Chinese postdoc met me at the door to the imaging lab and walked me back to the MicroPET setup.

He put me in a chair and continued what he was probably working on before Dr. Gambhir interrupted him. Fine with me—I was happy just to watch and learn.

There was another Tupperware with ugly pink hairless mice again, this time under a bright light.

"Andy, right?" Without looking at me, he pulled one of the nasty hairless creatures out of its container by the tail and shoved it into a Plexiglas cylinder, with its tail running down a slot in the cylinder and then sticking out a hole in the bottom.

"Yes, and you are . . ." I started trying to pronounce his name. "K-s-Y-a-o-y-u-o . . ."

"Call me Shawn," he said.

"Thanks, Shawn." Whew. I kept watching. He shoved in a white cylindrical plug, up against the head and nose of the mouse.

"Mouse restrainer," Shawn explained.

"Oh" was all I could say.

Out came a syringe. Shawn walked over to a pile of lead bricks and started filling the syringe from a vial encased in a big metal container.

"Copper-64. We get it from a lab in Wisconsin. They FedEx it. It loses two half-lives on the trip, but that's okay, it lasts plenty long." Shawn talked while he carefully pushed the needle into nude Mickey's tail. I'm pretty sure I heard a scream, but that may have been me. Tiny mouse turds filled the bottom of the Plexiglas container.

"We heat them up to open up the veins. Sometimes they're hard to find, but you get pretty good. I'm just a chemist, but here I am sticking needles into both veins in mice tails. Oh, well. These nude mice are easier—you can see skin. With black mice you just jab away."

"I'll keep that in mind." I laughed. Defense mechanism.

Shawn dropped the mouse into a knock-out box and within a minute or so it had stopped twitching.

"This mouse is completely immunodeficient. Nude mouse. We injected human breast cancer cells into the left mammary fat pad."

"When was this?"

"About a month ago. There should be a tumor in there. Now we want to functionally image it, not just use a CT to see it. So we send in a probe."

"That finds just the cancer cells?" I asked.

"We hope so. Yeah, with PET, the only approved probe for humans is FDG, you know, sugar with fluorine-18."

"I've read about that."

"Yeah, it's okay for some things, but worthless for prostate or brain cancer. We think we have another one—RGD peptide. The arginine-glycine-aspartate sequence is involved in cell adhesion. You know about integrins?" Shawn asked while loading the knocked-out mouse onto a tiny bed on the PET machine and attaching a hose to its nose to keep the anesthetic flowing.

"No," I said. Shawn set the scan time to five minutes.

"Angiogenesis?"

"No," I admitted. I was feeling dumb again. This was getting more common.

"Hmm. Okay, real simple. We have some time. Angiogenesis is the formation of new blood vessels from existing ones. Tumors and cancer cells are constantly creating new vessels, so we try to target that. There's a bunch of molecules involved, vascular endothelial growth factors, VEGF, and epidermal growth factors, EGF, and other things.

"There are drugs that target those, right?" I was trying to up my smarts quotient.

"Very good. A bunch of drugs. But then there are these integrins—adhesion molecules. This one specific integrin, alphavbeta3, is specifically expressed on proliferating endothelial cells and tumor cells that are forming new blood vessels. That's how we catch them, we find these integrins."

"So the RGD finds these integrin molecules?"

"Pretty specifically. We target this alphavbeta3. That's what the RGB peptide does. It finds it and sort of attaches to it."

"And the copper? Why not fluorine?" I asked.

"Let's see if I can explain. We attach something radioactive to the RGD probes so the MicroPET can light up the tumor. Let me show you some old images while we wait for this one to finish." He clicked around. "Where is that thing? Maybe this one . . ." I couldn't quite make out what the image was. You could sort of make out the shape of a mouse, but there were red splotches and streaks of yellow and blue blurs. I squinted my eyes and probably shook my head.

Shawn started laughing. "Oh yeah, not a good one. That mouse woke up in the middle of the scan and started running around." He laughed louder and harder. I started looking around to see if they had a broom for these things.

"Okay, okay. We use radioactive, but we can probably attach something optical, too, something that fluoresces. Mice are pretty thin skinned, so it would work with them, but not really for humans. Too much fat. Anyway, fluorine-18 is just for imaging. The nice thing about copper-64 is that maybe we can use it for imaging and therapeutics."

"Meaning what?" I asked.

"Well, we only use 17.4% of the radioactive payload for

imaging. That's enough to get the PET scanner to see it. But we can increase the radioactivity, put in a bigger payload."

"To do what?"

"Pretty focused radiation, right? If RGD peptide is good at finding tumors, then you can just up the radioactivity and kill the cells, not just see them. Pretty good, right? These PET scanners are pretty good. Brain surgeons use them in surgery."

"To guide them?" I asked.

"Well, actually they do a CT, MRI and PET scan of the brain beforehand and then tape up a printout of the scans on the wall. It's not real time. Could be."

The five minutes were up. I was watching the timer count down from 300 seconds.

"Okay, let's see what we've got." Shawn clicked away and three separate images popped onto the LCD display.

"Okay, this middle one is top down, back to belly. The other is back to front and then sideways. We use the FBP algorithm, filter back projection—you know about that?"

Ah, Bracewell shows up. I knew he would.

"Okay, let me set the brightness and contrast of scans and . . ."

I could see it. There on the upper left of the mouse outline was a circular red area. I pointed. "That's the tumor, right?"

"You've got it. It's not a very big one. A regular CT scan would have missed it. Pretty clear, isn't it?" Shawn kept adjusting the colors until the red blob was crystal clear.

"I'd say it's pretty black and white, but it's red on blue." I didn't need 12 years of med school to find that tumor.

I was staring at breast cancer tumor, but in a very different way than I had in the clinic on that Saturday morning with Dr. Solon Finkelstein with black-and-white X-ray films. There, he was staring and squinting and holding up a magnifying glass and racking his brain for patterns to justify labeling what we were looking at as possible cancer.

All of a sudden, I thought I heard bells going off—ding, ding, ding. I was looking inside a mouse, using Bracewell's algorithm, and there was a huge arrow, a giant flashing neon sign that was screaming CANCER HERE! You couldn't miss it. The signal was coming from the inside of the mouse heading out. No backup reading, no computer-aided detection from R2 or anyone else. There it was, ugly breast cancer, staring back at me with a snarl, begging to be zapped or sliced with a scalpel and tossed into a red bag.

There it was, my eureka moment number four. I hadn't had one in a while and was getting pretty antsy. You can locate cancer cells by injecting molecules that go find them and light them up. Perhaps early detection isn't all that far off.

"This molecular imaging has more uses than you think. We can image as I just showed you, we maybe can deliver radiation, too. But drug companies can also check their compounds with this. For example, you saw the kidneys light up. I'd say most companies don't even know that their drugs went there. Now they can see it."

"And how does that help?" I asked.

"Well, people take drugs to fight tumors, but we can only guess at side effects. You know Taxol? For breast cancer?"

"I've heard of it." I thought I had, anyway. From the conference? Buy Phase II, short Phase III.

"Very famous, works on early-stage tumors. Very successful. From the yew tree, I think, although now they synthesize it. Anyway, only one tenth of 1% of the drug goes to the tumor. The rest is scattered around the body. Now we can see where it goes. GE is doing a lot of work with angiogenesis and imaging. Merck has an RGD for drug delivery, but different from our probe. There's a lot going on," Shawn told me.

He kept moving sliders around, adjusting colors until two

big yellow circles and another duller yellow splotch showed up in the middle of the mouse image.

"Those aren't tumors, are they?" I asked.

"Kidney, kidney, liver," Shawn pointed. "The excess probe has to go somewhere. We could clear it out by squeezing the bladder to evacuate them, and—"

"Yeah, that's okay, I get the idea," I said and waved my hands. I think I wet my own pants, I was so excited. Another batch of doctors just got pink slips.

40 million mammograms done every year. That's 160 million films. Over 100 million of these files are read by not one but two doctors, and the only thing they can identify are growths and clumps of calcium that are visible on Kodak black-and-white film. Debates still rage whether mammograms are effective, let alone cost effective. Even with computer-aided detection from R2 and others, what a complete and udder, I mean utter, waste. Every year, a couple of billion dollars is spent on mammograms, plus all the mental anguish of false positives and risk from biopsies, all to find 200,000 cases of breast cancer. That's easily $100,000 per case, just to find it.

Mousey-Mouse had just showed me another way. Instead of shining a flashlight and looking for shadows, put the flashlight inside and shine the light on the inside heading out. Can it be cheap? Of course it can. You get those radiologists out of the loop. Mammograms become a product instead of a service. "Here is your image attached to your email. See any red? Nope? Good, you're clean, come back next year."

You can almost guarantee there is a decade of hassles and infighting and testing before this is ever used on humans. But I've seen the light.

Part VI

Chapter 43

■

Silicon and Biology

George, what can we do to shake this place up?" I asked.
"Depends what you want to talk about."

I was asked to introduce George Gilder to the crowd at Tony Perkins's AlwaysOn conference. It's a great conference, mainly about where the Internet and the Web are going, about blogs and wikis and the like. I always hate me-too kind of talks, and I think Tony set George and me up as a kind of freak show to talk about something completely different. I didn't want to disappoint. We had about 30 minutes before we went on, so we huddled outside and chugged coffee until we came up with something.

"It's got to be something a bit out there," I said.

"I'll talk about anything. What have you been working on lately?" George asked.

"Well, I've been playing around in health care," I admitted.

"You mean medical records and stuff like that, the Internet as the hospital IT department?" George asked.

"Not really. That's probably neat stuff, but I'm trying to figure out if health care has any of the same characteristics as Silicon Valley."

"You mean Moore's law? Metcalfe's law? Mead's law?" George asked.

"Maybe all of them. Things that scale, elasticity, three-year product turnarounds, new markets that get created overnight. That kind of stuff."

"Okay, okay. I get it. You'd think it's a natural as we get to smaller and smaller chip dimensions—we can digitize anything. But what do you think really happens when chips go nanotech, you know, nanometer dimensions?"

"Isn't that what Intel did ten years ago in their fabs?"

"You got it—about 10 years ago, they got the gate oxide below 10 nanometers; that's nanotech by any standard. Nanotech is really an extension of Moore's law. It's the assumption that things will continue to get better as they get smaller on into the next century and that this continued descent into the pits of the quantum will yield ever more exciting and valuable technologies."

"I think it probably will, for 20, 30 years, anyway," I said.

"Sure. That's what Ray Kurzweil is saying, about a singularity coming. He's got a terrific presentation of this argument and it's got lots of insights about the whole range of technology. But it is fundamentally based, I believe, on Richard Feynman's mistake way back in 1957 or so when he gave his famous speech 'There's a lot of room at the bottom.' "

"Mistake?" I asked.

"Incidentally, I discovered that the real author of Moore's law and any law of accelerating returns is Henry Adams."

"Who?" I had never heard of him. George always finds these things.

"Henry Adams wrote a book *The Education of Henry Adams,* where he propounded the law of accretion of progress. Sure enough, in this book there is one of these exponential graphs about the doubling index of worldwide energy consumption, which he took as a proxy for the advance of civilizations and technology. So it's been going on a long time and I think all these projections are based on a view of a kind of potentiary science resembling the end of the nineteenth century, when they thought all the real inventions had been exhausted and all the physics was known and then the quantum revolution came and threw this into a cat's hat and people still have not begun to come to terms with the real meaning of this quantum revolution."

"Okay." This was getting weird, but I suppose that could work.

"We all think that quantum theory means things always get smaller, there's room at the bottom. But in reality, quantum theory makes things get bigger rather than smaller. That quantum effects are not restricted to atoms as people believe but rather reach out across the universe and have a fascinating interplay with thermodynamic effects."

"Reach out across the—" My interruption was ignored.

"But going back, Feynman's speech really envisioned nanotech. He really defined nanotech as Eric Drexler and Kurzweil and all the others acknowledge and it's the atom-by-atom bottom-up construction of molecular machines. And this essential claim is physics über alles—physics prevails and the reduction of chemistry and biology to physics."

"All things get digitized?" I wondered.

"Maybe not. It all could be wrong. A mathematician at IBM named Gregory Chaitin, who is a follower of Kurt Gödel and Gödel's incompleteness theorem and Alonzo Church—all of these people who have shown essentially that Feynman was wrong about the ability to subsume all other science essentially under physics."

"I think we got them with the incompleteness theorem." George again ignored me—he was on a roll.

"Chaitin calculated the complexity—the information content of physical laws—by reducing them to lines of software code and found the complexity, the information content of physical laws, is radically exponentially smaller than the content of biological laws and biological phenomena and biological information."

"Okay?" I stopped interrupting. This was going somewhere.

"So it's impossible to reduce in principle—mathematically impossible to reduce biology to physics. And a professor here at Stanford, Robert Laughlin, also has demonstrated this. He's a Nobel laureate and he explained the fractional quantum hall effect—the crucial semiconductor insight—and he too shows that there is less and less at the bottom. When you go beyond the levels that we've currently reached, it turns out that nature operates emergently and on a larger scale than nanotech. Things don't get smaller, they get bigger. But that's a good thing."

"You mean we've reached the point already where electronics and biology work together?" I asked. I still wasn't sure where George was heading. This was nothing new.

"That's pretty much it. You can't theorize it—you just have to do it. Carver Mead has written a book about this called *Collective Electrodynamics*. And the basic principle is: If you can build it, you can understand it. And that is why engineers—why Silicon Valley—out-innovates the academic scientists. They actually build things. I wrote in my book *Silicon Eye* the story of what happened when a team at Cal Tech led by Carver Mead attempted to build an eye—the essential electronics of the eye in silicon—and discovered that academia couldn't offer anywhere near a satisfactory explanation. Academic biology could not offer a satisfactory explanation of the eye of a fly. And so how the fly can elude the swatter is beyond the real insights of biology. So they had to build the eye and improvise themselves and they came up with a radically new imager with fundamental inventions that allowed them to use the frequency-dependent absorption of silicon of light to build an imager that collects all the light at every pixel and reproduces in essence the way the human eye's cones collect light."

"So you don't know until you put it together and try it?" I summarized.

"Carver has another company, Impinge, with a device a few square millimeters—essentially a grain of sand that is powered by the incident radiation from the reader, which can be 40, 50 feet away. All by combining analog and digital and radio frequency all on this grain of sand and power conversion and a couple kilobytes of floating gate memory and no pins obviously. And it has to sell, not for $700 but for four to eight cents apiece. And this is a kind of real innovation that reflects the ingenuity that can be unleashed by effort to mimic biology."

"Mimic biology. If you can build it, you can understand it." I tried to summarize again.

"This is really the most fertile area today—the interplay of biology with silicon. There are major kinds of innovations that are aligned with real physics and biology and chemistry, rather than attempting a futile effort to pursue the quantum inspiration deep into the pits of uncertainty where the apparatus is the essence of the invention."

"I think we've got ourselves something to shake them up!" I said. "Let's roll."

Perhaps electrons and biology were going to collide sooner than I thought.

Chapter 44

■

To the Hutch

I got an email from Don Listwin:

> To: Andy Kessler
> From: Don Listwin
> Subject: Hutch
>
> I'm heading up to the Hutch on Tuesday morning. If you want to join me, I'll introduce to Lee and some of the other folks.
> Don

I responded almost immediately.

> To: Don Listwin
> From: Andy Kessler
> Subject: Re: Hutch
>
> i've rearranged my tuesday so it turns out that now i am free, and can join you on your trip to the hutch. Just let me know what flight you are on and i'll book tickets today. it may take a day or two, but i'll have to find my tie in the back of my closet.
> andy

The next day, I got this note:

To: Andy Kessler
From: Don Listwin
Subject: Re: Re: Hutch

Tail# N725CC out of San Jose. 9 a.m.
Leave your tie at home (if you even have one)
Don

I got it. This was no Alaska Airlines to SeaTac Airport.

The ride to Seattle in a Citation was uneventful. It wasn't Listwin's plane—he needed to lease his own out for a bunch of time to qualify for some tax break, so he was leasing someone else's. All too confusing for me.

I'm not a nervous flier, but these private jets make me a little jittery. It was a beautiful day—the ride was smooth, the small talk friendly with Don and his wife, Hilary. I got a rundown on some of the folks we were going to meet with. Mostly I read the paper and hid my annoyance that there were no little bags of pretzels or peanuts and that I was facing backward. I could deal.

There was a little fog over Mount Saint Helens, or was that steam showing that it was about to erupt? These small jets can't handle a volcanic eruption, can they?

I decided to ask a question about something I might understand the answer to.

"Don, you keep talking about this bioinformatics thing. Is it just some Oracle database? Is there something special about it?"

Don leaned forward. "Nothing is easy in this world. We need a database of all known proteins plus a standardized way to store information from new proteins as they are characterized. The National Cancer Institute has something called

CaBIG, the Cancer Bioinformatics Information Grid. It tries to define things from clinical trial data to genomics—even biospecimens."

"Okay."

"We'll be part of it. Canary Fund is giving half a million to the Hutch to develop a proteomic database. It's Computational Proteomics Assessment System, CPASS. Maybe we'll check out the mass spec lab and you can see the output and it'll make more sense."

I looked out the window. We were coming into Boeing Field. I've driven by there tons of times, after landing in SeaTac on the long taxi ride into Seattle. Cool. This was convenient, probably a five- or ten-minute drive into town.

Don went on about an informational network effect and biomarker bank but I was enjoying the approach too much to pay attention anymore.

We hit the runway a little hard, but no big deal, that happens on 757s all the time. No problem, we were wheels down on the runway, and Don was going on about a consortium with a Korean institute and their analytics when a loud *boom* came from under the jet.

"What was that?" Don asked. The three of us looked around. It didn't seem like a problem but the jet came to a quick stop sort of halfway on and halfway off the runway. No smoke or flames. We were stopped. But my heart was beating pretty quickly.

"What was that?" Don asked the pilot when he stuck his head out.

"Not sure yet, sir."

"Is this something I should worry about?" Don asked.

"I think it's just a blown tire."

"Let me see." Don got up and moved forward.

The pilot opened the door and Don was the first one out

on the runway to check out what was going on. I watched him and the pilot look around, and figured the last place I wanted to be was on a jet in the middle of an active runway, so Hilary and I got off, following Don and the pilot.

I looked across the runway and saw a huge garage door open and a green fire engine emerge with its cherry tops flashing and its siren blaring, and head toward us.

Don was behind the landing gear looking around.

"What's this?" he asked the pilot. There was a 15-yard skidmark behind the tire. The fire engine was almost there.

"Well, looks like the brake locked up," the pilot said in classic pilot drawl.

"So, this is something I should worry about," Don sighed.

The fireman got out wearing some funny hip waders, probably flame retardant, I didn't want to know. Jets were still landing on the other runway—ours seemed to have been shut down, except for what appeared to be another jet about to land. "Okay, we've got to get you folks off the field quickly," he said. "You can't walk across an active runway. You ever ridden in the back of a police car?"

"I have," I answered. "But I don't like to talk about it."

I got a funny look from Don and Hilary but a smile from the fireman.

Still, here we were, sprawled halfway on a major runway with big-ass jumbo jets queued up to land on the other runway, which was still active, and Don was poking around under the plane to see if he could fix it himself.

An SUV with cherry tops, and again, sirens blaring, pulled up and a woman in uniform got out to check out what was going on. "Don't worry about that other jet over there, it's only using a short runway, won't get close." It screeched overhead—close enough for me.

I wondered if this whole episode was staged by Don Listwin as a metaphor for how chaotic and dangerous the waters of medicine can be and that the best-laid plans of early detection can sometimes leave you sprawled in the middle of active runways depending on emergency workers to escort you off.

Chapter 45

■

Mass Spec

The Hutch was nestled up against I-5, just north of downtown Seattle—close enough to smell the coffee. Its buildings looked no different from those of any industrial park in Silicon Valley. You certainly couldn't tell from the street that this was a leading center for cancer research.

"Don's got a board meeting for about another hour. We can grab some lunch. You guys were a little detained?"

"Mechanical problems," I offered.

"Let me make a few calls to move things around." Pat McCowan was my guide for the day. She used to work at the Hutch, but Don Listwin hired her to work at the Canary Fund.

"Sure, that'd be great," I said.

"Here comes Marty McIntosh. You'll be meeting with him later, but why don't I introduce you and you two can chat for a second while I try to find Sam."

"Hi, Marty McIntosh. You with Don?" Marty had a big smile and straight black and gray hair flying all over the place, like he had just gotten off a motorcycle.

"Yeah, just trying to get up to speed quickly on some of the stuff going on here. You are the guy doing the protein database?" I asked.

"Yup. That Don is something."

"What do you mean?" I asked.

"Well, lots of people give money to the Hutch and other places, for that matter."

"Yeah?" I wasn't sure where he was going.

"The perfect giver is one who writes a check and then goes away—until they're asked for more. Don's the kind of guy who gives money and then shows up and tells you how to spend it."

"I don't know him that well, but he seems like the type," I said.

"Don't get me wrong, I think it's great. He's got a pretty strong opinion and definite goals of what and when to get stuff done. It shakes places like this up. He's single-handedly gotten early detection high on the agenda around here. I wish there were more guys like him."

"Okay, Sam will be back in after lunch. We'll check out the mass spec lab, but first we have time for a quick bite. You like salmon?" Pat asked.

"Have you seen a mass spectrometer before?"

"Actually, I've seen one in a physics lab somewhere along the way." I knew they have been around for decades.

We were in some basement lab in the bowels of the Hutch. A box the size of a large microwave oven was sitting on the lab bench, a thin tube coming out the front and some electrical connectors out the back, connected to a PC.

"So you know what it does?" the lab technician asked.

"Sort of," I mumbled.

"That's right, it sorts by molecular weight."

"Explain what we use it for," Don Listwin piped in.

Don was back from his board meeting and was probably wound up after sitting with 50 other benefactors listening to a few too many PowerPoint presentations.

"Well, we're interested in identifying specific proteins in tumors. Each protein, fortunately, has a unique molecular mass. So we run samples through these mass spec machines and try to find these proteins that might end up as biomarkers."

"So what's this tube?" I asked. I was looking for open containers or petri dishes but didn't see any. Wait a second, I started thinking—this was a cancer center. I started breathing in shallow breaths and covering my mouth with my hands. I hoped I wasn't too obvious.

"Well, you can't just weigh molecules. What this machine does is ionize them, turn them from solids and liquids into gases. These ions carry a charge, think of them as an electron spray. That spray is blasted toward the back wall. Slow-moving particles crash on one wall and the lighter, faster ones crash on the other wall. You adjust the magnetic field inside the machine to do an analysis in some specific range of mass, which hit the detector on the back wall."

"But there's lots of stuff in each sample," I said.

"Sure, but you get a chart when you're done. It's a histogram, you know, bar chart, of how many particles of each molecular weight are in the sample."

"How much is one of these things?" I asked.

"Probably 300 thou, plus or minus."

I whistled.

"I think I paid for this one," Don said as he squinted to looking for a serial number or something.

This whole setup reminded me of the early days of the semiconductor industry. Today, there are billion-dollar fabs with hardly any people in them, but in the early days there were ovens that would heat material, and researchers would peer into them occasionally to see if what they were making was done so they could go onto the next step. Once perfected, it would be used to make parts, and then the next later and greater process was again perfected by trial and error.

"And it's a one-off thing? Can you do measurements in parallel?" I asked.

"I heard about a guy who had six of these running in parallel."

"Not eight, not four, not sixteen. Just six?" Don asked.

"I think it had to do with the footprint of the box and how many detectors he could cram in there."

The learning curve was starting.

While he was answering, I took a look at the chart on the screen. Sure enough, there were blips at certain molecular weights along the axis. On the bottom of the chart was a footnote that read "Total analysis time: 30 minutes."

"30 minutes?" I asked, looking at Don. I was used to fast Pentiums and gigabit data rates. "I'd have guessed 30 nanoseconds."

"Biology is slow."

"Okay, you've had enough of these mass specs. Let's go see Sam," Don declared.

We started walking through corridors, went up elevators, through walkways and then more of the same. This was a huge complex.

"Sam's the top early detection guy around," Don said about three or four times.

We wound our way through the maze of corridors and elevators and finally walked down a long hallway, when Don stopped suddenly and motioned me to follow him into a lab. Lots of test tubes and beakers and what?—I tried to recall my freshman biology—Bunsen burners, maybe? The lab was huge. There must have been a couple of dozen folks in lab coats milling about, typing into keyboards and drinking coffee. I breathed a little easier. If they were drinking and munching on Cheetos, I supposed I could take a deep breath.

"See this place?"

"Pretty big lab," I said.

"This wasn't cheap, but we wanted Sam. He's the top biomarker guy around. Was at U of Michigan, and we brought him over here. Not that Ann Arbor versus Seattle is that tough of a choice. But he insisted on duplicating his entire lab. I got the bill, more or less. I think it will be worth it. He's got a brute force method to identifying proteins that could be biomarkers, but it works."

We left the lab and walked around the corner and into an office with the most beautiful view of Lake Union, with planes taking off and landing on the water. I guess they really did want this guy—he got the best office in the building.

I looked around his office for clues about him. About all I could make out was an M.D. degree from the American University in Beirut and another from the University of Michigan, who I usually root against, in football anyway, well—basketball, too.

"So, what can I do for you?" Dr. Hanash asked.

"Well"—gulp—"I'm from Silicon Valley . . ."

"Like Don," Sam said.

"Yeah, anyway, I'm a tech guy and most things that I look at go down in cost every year, and I'm trying to see if there is any part of medicine that does the same thing . . ."

Sam let out a very hearty laugh. I get a lot of that when trying to explain what I'm looking for.

"And everything in our world goes up in cost," Sam said.

"Seems like it."

"But you're right. There is no doubt in my mind that with new discoveries, we will be cutting costs. But discovery is rather expensive and disjointed. Who's going to pay the costs so we get the benefits? The private sector. They're not willing—"

"Not yet," Don interjected.

"The government is not organized around discovery, not in early detection. It's just ideas, no master plan, no road map," Sam complained.

Another plane landed on the water. No brakes to lock up on those, were there?

"Breaking the conundrum is a challenge," Sam said.

"But who pays for all this research now?" I asked.

Don started fidgeting. I think he was paying.

"NCI. But the amounts are tiny relative to the whole budget," Sam told me.

I didn't know the acronym.

"The National Cancer Institute gets its funding from the National Institutes of Health, which is taxpayer dollars," Don said for my edification.

"So it's political how the money is allocated?" I asked.

I wish I hadn't. I got a look from everyone in the room that said, "Of course, you naive puppy!"

"But it's just research," Sam went on, "that's the problem. Early detection crosses over lots of disciplines, chemists, biologists, physicists, computer guys, and so on. If I ask someone to work on a project and we file a paper in a year or so, and this person's name is number 20 on the paper, when he goes looking for a job and they ask what research he's done and all he has to show is number 20 on some paper, that's not good enough. It ruins a person's career to work on these things. It's crazy. I feel bad asking someone for help when they could be advancing their career in their field. There has to be a better way to create incentives for 50 people."

"Isn't that what stock options are for?" I threw in.

Don laughed. "Wrong crowd, but you're probably right. Even a big company like Cisco knew how to incent groups of 50."

"There's a need to industrialize research—smaller organized projects. The mind-set needs to change, and then we get a road map. Pressure from guys like Don—and of course, a working prototype. There's nothing like something that works to get a road map."

"Explain what you do, Sam," Don prompted.

"I've been obsessed with protein's relation to disease for 20 years. It used to be hard to justify proteomic research, now it seems to be the in thing. I don't worry about time or cost, but creating a strategy that works."

Refreshing honesty!

"There are millions of proteins, probably 10,000 we care about. We have to painstakingly find these proteins. You find the unique proteins from cancer tumors and you can screen everyone for the disease. We have to create a mini-industry, a production line. When it works, others can copy it and make their own discoveries. Plus, this place has some neat stuff," Sam said as he gazed out the window. I thought he was talking about the planes.

"A number of years ago, there was a woman director of the NIH who spearheaded the Women's Health Initiative. A billion dollars directed out of the Hutch. 160,000 women were seen and blood samples were taken on every visit and archived. The program was investigating estrogen replacement therapy, which, er, was proved nonbeneficial. But we have the source material to look for proteins for almost every type of cancer that some of these women may have ended up with, well, except prostate cancer, for obvious reasons."

"Then what?"

"Right now we can do eight samples a day. We have to get through 10,000 or more, you figure out how long that's going to take even if everything goes perfectly, which it never does. We need more machines, we need automation, we need informatics to archive what we learn."

"That's the CPASS I was telling you about," Don added.

I nodded.

"We have to do this systematically, or else we're just playing roulette. And then we create the antigen for assays. You have to narrow the proteins down just right, because antigens are expensive to create and it can take a year or longer. Then we have to prove that a panel looking for these proteins actually works as a real cancer screen."

He took a deep breath as he pondered the Mount Everest of a task he had just laid out for me.

"But we'll get there. We know what we have to do. It'll work."

With that he nodded his head definitively.

"Oh, and one more thing. When it works, when we can do early detection on everybody—for cheap—it turns medicine inside out."

I'd been waiting for someone to say that.

"What was the antigen thing he was talking about?" I asked Don.

"You know about ELISAs?" he asked me.

"The pregnancy test thing?" I asked.

"Once you have a protein you think is a candidate for a biomarker, you can create a pretty easy chemical test to look for it.

"First you have to create the antibody. So you inject the protein into nude mice . . ."

"Those pink hairless ones. They're a bit creepy."

"Yeah, those."

I knew the rest—harvest antibodies from mice spleens—monoclonal antibodies and all that.

"It can take close to a year and lots of money—50 grand or more. You want to make sure you have the right protein."

"I can see that," I said.

"Someone around here thinks they can create the same antibodies out of yeast—in two weeks instead of months or a year. Cheaper, faster—and it gets PETA off your back."

Chapter 46

∎

Hartwell

Don ran off to make a phone call, so Pat walked me through some more labs on our way to meet with the guy who runs this place. We stopped next to a plaque honoring major donors to the Hutch. Paul Allen, Bill and Melinda Gates, Jamie and Karen Moyer Foundation—I think I saw him pitch once for the Mariners—and the Listwin Family Foundation.

"Oh, here's someone you should meet," Pat said. "Nancy, this is Andy Kessler. Nancy is in charge of development at the Hutch." I always run away from anyone who has development in their title. They look you in the eye while they feel around for your wallet.

"Hi, nice to meet you," Nancy said. I unconsciously put my hands in my back pockets. She turned to talk in a somewhat lower voice to Pat. "I got the check for 500, uh, units from Don. Thanks."

"Oh, good," Pat said.

"I know it's for the CPASS stuff, but I had him leave that off the check. It's directed to our general fund. This whole over-head thing is a pain—I think he insisted it all go to this database thing."

"That's probably best," Pat said.

I don't think I was supposed to hear this conversation. It

sounded like gift arbitrage, just like plane-leasing arbitrage, I suppose. You lease someone else's plane and lease your own out, because of some strange tax rule. I'm not sure I ever wanted to know how a foundation shuffles money around, but it sounded like Don Listwin was a quick study.

Looking over at me, Pat said, "We should probably go meet Lee. Bye, Nancy." Turning to me, "Now, you know Lee is the president and director of the entire cancer center. He won the Nobel Prize for medicine in 2001."

Gulp. Well, I had met with Arno Penzias once. He was playing with noise in his antennas at Bell Labs and postulated it was the echo from the Big Bang. I suspect this prize was for something more useful. "Lee worked with single cell yeast, baker's yeast actually, and found that certain genes controlled normal cell division. These same 'checkpoint' genes are in humans, too. It changed the way cancer is thought about, quite remarkable, actually," Pat said.

Great. Here's an international hero, and I've really got nothing to ask him about except if he knows about any technology that might somehow help. I think the whole baker's yeast thing had thrown me off.

I stood up at Dr. Hartwell's whiteboard and drew a simple graph. An X axis, a Y axis and a downward sloping line that dropped from left to right. It's all I had—a Rorschach test for anyone I met outside of technology.

"This is how Silicon Valley works—costs go down every year and create new markets."

Lee looked over at Don Listwin. "Is that how all you people down there think?"

"Something's in the water." Don chuckled.

"Well, could be, could be. We don't have those kinds of curves. Although"—he paused—"it does remind me of something."

He got up and grabbed the marker. He wrote "1 million compounds" at the top of my sloping line, on the upper left. Then he wrote "1" on the lower right, at the bottom of the line.

"That's the drug discovery business. Has nothing to do with cost, quite the opposite, but we put a million potential ones in and get one out, and the one coming out isn't even all that good," Lee told me.

He let that sink in for a while. I think he was talking about the funnel.

"There is a huge promise from this molecular biology revolution. It's just different from what most people think. It's not some magic bullet—but rather moving earlier in health management."

"You mean cheap, mass market screening?" I asked.

"I certainly hope so." I was hoping for something more definitive. There had to be some money quote from a Nobel laureate.

"It's a paradigm shift. That's what you techies like to say. It's all therapeutic or devices—very little in diagnostics. But it will be just like IT, very primitive right now, so hard to recognize, but we can already see what we can do, so it will happen," Lee said vaguely.

Again he paused. I'm not sure if he was letting it sink in or gathering his thoughts for the next blast.

"This is a revolution in health care. Not just cost, but prevention versus acute care. You know how important that is?"

"I think so. Is this over the next five years?" I tried to pin him down.

"Hard to put a time on it, we need advances in research technology to get there—more information out of molecules, RNA, DNA. More advanced tools. We know how to do it with today's equipment. We need to get the toolmakers to move with us." This again sounded like Silicon Valley.

"We know how complicated this is, we have made incredible advances in just a few years of research. I'll tell you, though—we may have to go international to test out some of these advances. The rules on human testing are a little tight."

I might volunteer.

"You know about PET imaging?" Lee asked.

"I do," I said.

"We can do imaging with FDG, radioactive-tagged glucose. It's been available for 30 years, but only just recently approved for reimbursement and for use in clinical trials. Can you believe that? We're terrible at implementing advances." Dr. Hartwell's face was starting to turn a little crimson.

"The whole system is crazy. What Don and I are pushing is a proof of principle." Don was nodding his head. Nice to be given credit by a Nobel. I guess if you're paying for it and pushing it . . .

"It's pretty straightforward. We have to be able to accurately measure thousands of proteins, and find the ones with strong predictive values—from serum, from tissue or from urine, it doesn't matter. The proof is the predictive value. We get that, we can roll it out."

"Bleed to read, right?" I asked.

"Exactly," Lee said. It was nice to know the lingo.

"And this can be a first screen?" I asked.

"Someday. First, you lower false positives by narrowing down to the high risk, familial risk, genetic risk and then this phenotype risk. False positives are expensive, way more than the cost of the tests. It's the test, and then all the back end if you show positive—scans or biopsies, whatever."

No different from what John Simpson told me. Only when the scans were reliable could they become commonplace, because the back end was expensive.

"So you are confident?" I asked. I wanted to see him either turn red or blurt out the secret to what they were doing.

"Let me answer it this way. Science is something you don't

know the answer to. We know the answer here. It's a panel of markers with 99% sensitivity and 99% specificity. So it's not science, it's an engineering problem."

Wow, I wondered if anyone else in medicine thought this way. "And technology is how we get there."

He got up and drew another chart. It was a straight horizontal line. "See that, that's the cost. Whether we get one biomarker or 20 or 50, it pretty much costs the same. Once the methodology is set, we can put thousands of biomarkers through and it becomes cheaper per biomarker. That's probably somewhat similar to what you guys do down there. There's no silver bullet, just lots of copper bullets."

"I like that analogy," I said.

Rightly ignoring me, Dr. Hartwell picked up steam. "We just have to show that screening reduces mortality. It's the ultimate test. It's been done before—Pap smears meant 80% to 90% reduction in cervical cancer; colonoscopies meant 70% five-year survival rates."

"Mammograms?" I asked.

"That's still controversial, the conclusions are iffy. But our screen should change that, of course. Why do X rays when you can look from the inside out? We just have to shift the curve, become less reliant on drugs. Early intervention has a high hurdle, but it's the right way to go."

"So you think cancer can become a thing of the past?" I asked, hoping for a simple yes.

"Look, we've got to prove it out for each cancer. There's no reason to think it won't work—we've got to try it."

He paused and then looked me in the eye for emphasis. I get flustered when people do that to me, but I stared right back at Dr. Hartwell.

"Drugs certainly aren't working. This really is our only option."

Chapter 47

■

Markers

I'm not sure we're going to get much of Nicole Urban's time. She's the one who's going to come up with the prototype for ovarian. That's what I'm funding," Don said. He had his money all over this place.

"She pulls that off," he continued, "then we have a platform for others to emulate, it won't be hard to attract funding."

"What do you mean by prototype?" I asked.

"An actual screening test that works. Today it's just CA-125 as a screen for ovarian cancer. The sensitivity of CA-125 is 84.7% for carcinoma of the ovary, fallopian tubes, and endometrium."

"Endo-what?"

"Lining of the uterus. 84.7% means that it misses 15.3% of women who actually have ovarian cancer. You can't recover from a false negative like that. So Nicole is running multi-antibody assays. The trick is to find something like 20 markers and get the sensitivity to 95% and then 99% and then early detection becomes reality."

We kept walking down the hall. I'd long ago lost track of where we actually were, still in Seattle, I assumed.

"Hi, Fred," Don said as he waved to a middle-aged gentleman drinking a coffee and deep in conversation with someone.

Fred looked up and said, "Oh, hi," and then returned to his chat. We kept walking.

"See that guy?" Don asked.

"Yeah?" I said.

"That's Fred Applebaum. If you ever get cancer or anyone you know gets cancer, he's the guy to talk to," Don said.

"Okay." That was about all I could come up with. I didn't plan on getting cancer and hoped to hell no one in my family got that sick. But even if they did, what were the chances I could get a Fred Applebaum to talk to me? "Hey, I passed you in the hall once at the Hutch" just didn't seem like a dialogue opener. As we walked and walked through the labyrinth of the Hutch, it struck me that what was really needed was a clone of Fred Applebaum to exist at every hospital, in every town. The guy's brain was probably a national treasure, yet only those lucky enough to buy him a cup of coffee could access it.

"Oh, wait, here's Nicole . . ." Don waved to a slight woman with wire-rim glasses and a white lab coat. She seemed reserved, maybe a tad shy, yet was getting flagged down in the halls of the Hutch by a gregarious Don Listwin.

"Hey, Nicole. What are we up to with HE4?"

"It looks pretty good."

"What's pretty good?"

"Point nine five."

"Really. That's fantastic." Don seemed elated. Hard to believe he could be more up than his normal self.

"It's getting there," Nicole said, trying to play things down.

"How about with mesothelin?" Don asked.

"I can't tell you yet."

"Because you don't know or because you don't want to tell me?"

"Because it's early, and I'm not sure of the results."

"But you have some?"

"Yes, but . . ."

"Aaaannndddd? C'mon. Let's hear it," Don insisted.

"About the same."

"As HE4?"

"Looks like it, but it's early."

"Nicole, this is big." Don almost jumped in the air.

"Could be, I've got somewhere to be, this ED seminar that's going on."

"You coming to dinner tomorrow night? I'm taking out the whole group, casual stuff, nothing formal, you should come. We can talk about this chip . . ."

"I'll try," Nicole said and then disappeared into a maze of conference rooms.

"I told you she'd pull it off. She may already have," Don said with a big smile.

"What did she do?" I asked. From what I could gather, she merged a Hemi engine with mescaline but didn't want to talk about it.

And something about a chip? I kept meaning to ask.

Meanwhile, I was on a biology thrill ride.

"I told you before—we think that if you mix CA-125 with other markers, the sensitivity goes up. Makes sense, you just have to pick the right markers and do the tests with utmost precision to prove it out. She's shown that CA-125 plus a marker named HE4 gets you to 95% sensitivity."

"That's what you're shooting for, isn't it?" I asked.

"For now, 99% is the goal, but just getting to 95% is really hard," Don said. "She's shown it for mesothelin, too, I gather. She'll work her way through others, probably do three-way assays to see what happens. Maybe she'll hit 98% and then we have a viable model." Don was beaming. I recognized the look—when someone's investment works, a stock pops or on the first day of an IPO.

Except Don's investment wasn't some return on capital. It was proving that blood tests with a set of biomarkers could accurately predict ovarian cancer and maybe tons of other cancers. I made a note not to check on any of my stocks that day—what was the point?

"So, who actually owns CA-125?" I asked.

"It's a Japanese company, Fujirebio. I think they bought the old Centocor Diagnostics. Maybe a couple of hundred-million-dollar businesses," Don answered, but I think I knocked down his elation.

"Aren't they going to be pissed off that you're finding other biomarkers that work?" I hadn't quite thought through the whole "who owns what" thing that drives the drug business.

"Tough question." There was a long pause.

"You don't know?" I asked.

"Well, it's complicated."

"Why?"

"Turns out that Fujirebio owns both HE4 and mesothelin, too. And the patents on them. We just proved in their markers. Ought to be a much bigger business for them," Don said.

"So then they'd be thrilled?" I asked.

"Probably. But Sam and others are finding other biomarkers. We're going to end up with 20 markers. And probably a rotating 20 as we prove new ones in. CA-125 probably stays, but you never know what we'll find out."

"And who owns the other 17?"

"Don't know yet. That's why I'm trying to create a protein bank. CPASS is where you store the information, but we need a mechanism to allocate funds to patented markers, based on contribution. It's not easy. That's my next big task—visit all these guys and make sure they play."

"I've seen this a few times before," I said.

"Where?" Don asked.

"That's how CDs got off the ground—Philips and Sony pooled patents and split royalties on the basic technology, and sold equipment, too. With DVDs, it was a bigger issue, more entities had patents that were needed. But they ended up pooling them and a committee somewhere decides how to allocate royalties. It works."

"I'll check it out," Don said.

I hadn't known him that long, but I was quite sure Don Listwin would figure this out. You don't get to a top spot at Cisco without knowing how to navigate through the red tape of bureaucracies. But corporate bureaucracies are probably the easiest, compared to patent lawyers and litigation.

"I've been meaning to ask—is there some sort of approval process for this stuff with the FDA?"

"We just draw blood and test it. There's an approval process, but it's pretty straightforward. You just have to show it works, not that it won't kill your patients, which is what drug companies have to go through. Here, you'll kill patients if you don't get this to market as quickly as possible. It's the imaging step that may be problematic, since you stick these tagged markers into people. But from what I understand, it's pretty innocuous, shouldn't be a problem. But those are famous last words, aren't they?"

"Let me introduce you to Marty, and then I've got to go," Don said.

"No problem, we met earlier. And I appreciate you setting all this up," I told Don.

"You're going to head back tonight? I'm more than happy to give you a lift back—on some other jet, of course—but not for another couple of days."

"I've got an Alaska Airlines flight." With working brakes, I didn't say. "But I'm sure I'll get strip-searched—happens every time I buy a one-way ticket."

Don chuckled and left me with the only person I had met

that day whom I could see sitting in a conference room in Santa Clara discussing routing schemes or wireless multipath problems. Marty McIntosh looked like a techie, from the flowing hair to his clothes.

"So you're trying to find things in medicine that go down in price?" he asked.

"Am I crazy?" I wondered.

"Probably. Some people think that anything that doesn't affect clinical care increases cost. It's kinda like, if you don't do early detection, then costs go up. But if you do do early detection, costs go up anyway." He laughed.

"I thought mammograms and Pap smears and that kind of screening stuff increased survival rates." I was confused.

"Yeah, they do, but they also increase cost."

"Really?"

"Sure, because the tests are overused. You can probably do them every five years, but doctors want them every year, patients insist on them, so the costs go up."

"But then people are healthier—so it's not so bad?" I asked.

"You didn't ask me about quality of life. You asked me about costs."

"Fair enough," I said.

"Look up Louis Russell. He wrote a book, something like *Is Prevention Better Than Cure?* A good question."

"Is it? I would think that it ought to be," I said.

"You'd think. People do spend money on preventative medicine, but then they drive fast or jaywalk. Maybe they really don't care about living all that long."

I thought I'd better talk about proteins, since I drove fast and jaywalked all the time.

"So what exactly is this CPASS?" I asked.

"Well, if you look at the Human Genome Project, it

worked because people shared data. In proteomics, we need to do the same thing, but proteins are more complex than DNA. It's lifetime employment for good scientists. Hurray for that."

"So it's a database of proteins?" I asked.

"Probably more than that. Don Listwin doesn't think small."

"I gather," I said.

"It's protein data and how to query it. Lists of proteins have no structure. We need all the mass spec data, molecular weights, everything that's part of protein sequencing."

"So it really is protein sequencing." I was trying to understand.

"What's needed is really more of a platform—a proteomics Google. You query for a protein from some disease. It's hopefully already there, or else you're working your mass spec and putting in the data yourself. So it's a little like open source. Use it or contribute to it. Linux for proteins. We've got some ex-Microsoft guys who built this thing and then can sell it to anyone doing research."

Linux for proteins? Now that's interesting.

"If nothing else, it ends duplication. Why should everybody be sequencing the same proteins? But it's not just that—it's like any open source project. Everyone gets smarter and gets what they need out of it, by only contributing in smaller pieces. That's the idea."

"And perhaps lower costs." I was going to flog this horse until it barked, or something like that.

"Yeah, I think costs are going to change, somewhere out there. If we can screen for these proteins on the cheap, we may yet break that 'costs go up because costs go up' thing. I hope so, anyway."

I played trip to Seattle backward now, except in coach. Sure enough, my one-way ticket got flagged by security, and I was stripped down, patted down, sniffed, swiped and checked for gunpowder fumes. But it was worth it. Like homeland security, early detection is going to unfold in fits and starts, money wasted and then finally a process to find cancer starting with simple blood tests. The only question was how simple and how soon.

As I dug into my molecule-size portion of peanuts on the flight back, I started getting pangs of doubt. Not that I didn't think that this would all happen, I was convinced it would. But I thought that I had just wasted my time.

Except for CPASS, I had just spent a long day learning about biology and chemistry. That wasn't all that bad, I suppose, but I was looking for something that resembled silicon— smaller, cheaper, faster, better.

Perhaps I found it in the front end of the business, finding markers. As Sam Hanash said, it's now an engineering problem. But that's just improving the productivity of research. That's not earth shattering—not earth shattering enough, anyway.

I was looking for something that could be delivered to you and me for cheaper and cheaper year in and year out.

Not to overthink this thing, but perhaps Marty McIntosh was right. Early detection saves lives, and it saves on chronic care, but only the expense of care for those who would have reached the chronic stage. If you didn't lower the costs of early detection, it would get overused and end up costing just as much or more than the actual care.

Those PSA tests and CA-125 tests are each $100. Figure a dozen different cancer tests for 300 million people in the U.S., spending $1,000 per year—that's $300 billion right there. I can't see that happening, not any time soon, not until cancer rates start dropping, which they will only if early detection screens are put in place—a nasty circle.

How much do these tests have to be to become earth shattering? $10? $1? A dime? I thought I was on the right path, anyway. It's the path of getting doctors out of the loop. Come to think of it, for most of us, doctors are already out of the loop. Their starting point is when we are already sick. Unless we are lucky enough to break our necks skiing and a scan discovers a tumor.

Where is the damn silicon? Without it, this is just a journey that ends up as a wreck spread across an active runway.

Chapter 48

■

NanoNano

I read that nanotech guru Richard Smalley recently died. Age 62. Of leukemia. At the MD Anderson Cancer Center in Houston.

Smalley was a chemistry professor at Rice University in Texas, but he also taught physics. Back in 1985, Harold Kroto of the University of Sussex was visiting Rice and talking to Smalley about wanting to understand more about carbon-rich stars. Is that what these guys do all day?

So Smalley and another professor, Robert Curl, enlisted a graduate student, Jim Heath, and set up an experiment using graphite—#2 pencils, perhaps—blasting it with a laser beam to see what happened. That might simulate the surface of these kinds of stars.

Digging into the results, they noticed lots of molecules made up of 60 carbon atoms. Instead of breaking down and turning into some other material, they stayed pretty stable.

Smalley went home, got out Tinkertoys or something and put together a structure of 60 carbon atoms, combining 12 pentagons and 20 hexagons that formed into something that looked like a soccer ball. A very small soccer ball. The structure reminded Smalley of geodesic domes—the ones that were always on the cover of *Popular Science* magazine as the futuristic structure we might all someday be living in, except we couldn't hang picture frames on the walls. Anyway, he called them Buckminsterfullerene balls, after, well, Buckminster Fuller, who came up with geodesic domes.

The rest of the world called them buckyballs. Much better name.

Buckyballs were strong and could carry electricity and because they were round, were kind of slippery. Kroto, Smalley and Curl split the 1996 Nobel Prize in Chemistry.

But in 1991, researchers at NEC in Japan had opened up the fullerene structure. They discovered that they could create an entire sheet of this honeycomb graphite and then roll it up. They ended up with a very long, very thin, but extremely strong yet flexible tube. Its dimensions were in the nanometers, so the tubes were named carbon nanotubes.

Zillions of new applications were thought up for nanotubes, mostly pie in the sky—like space elevators.

After my talk with George Gilder, I was intrigued. Maybe you could construct some really tiny robot that went into your body and zapped cancer or removed plaque. Sounds farfetched, and probably is. But I needed to understand nanotech a little better.

George W. Bush had recently signed a Nanotech Research Act. There was a National Nanotechnology Initiative. I figured it was a boondoggle, especially when I noticed that Mike Honda, a House member from Silicon Valley, was the cosponsor. $3.7 billion over four years—finally some pork for the Valley.

At the signing ceremony were bunch of folks I knew. Venture guys Steve Jurvetson—hey, maybe his table was available at Buck's—and Kleiner's Floyd Kvamme. I was going to track them down, but ended up running into another guy who was in the Oval Office—Josh Wolfe from Lux Capital. We found common ground—we both were born in Brooklyn and schooled high above Cayuga's waters.

"They slice, they dice? Do nanotubes do everything?" I asked.

"They will. Carbon nanotubes are a hundred times stronger than steel and a sixteenth their weight. They're more electrically conductive than copper and have better thermal conductivity than diamonds," Josh told me. I thought he'd used this description before.

"And?" I asked.

"And what? It's a materials science game. You can use this structure in building materials, paints, clothing, all sorts of stuff. It changes the economics of a lot of these industries," Josh said.

"And electronics? Intel and the chip guys have been doing nanotechnology for some time. You shrink and integrate for almost five decades and eventually you start playing around with some really tiny stuff. They're 65 nanometers now," I said. Nanu nanu, I thought.

"Sure, over time there will be some use. But semiconductors are already pretty remarkable. The trick is to make nanotubes in volume. Some of it is traditional stuff. You just have to try it."

I thought I'd heard this before.

"There is a company named Nantero that found it difficult to get carbon nanotubes to assemble in the right orientation, so they spin coated them on a wafer and then use lithography to pattern the new surface. That's pretty similar to what chip guys do. Still not easy."

"But you can almost use some of the existing equipment?" I asked.

"Sure. One of our portfolio companies, Molecular Imprints, uses e-beam lithography to create a mask, same way

Intel does it, but then they use imprint lithography to stamp out sub-50nm structures. Kind of like a wax seal from a sixteenth-century king—or maybe like the opening credits scene from *Charlie and the Chocolate Factory*. Stamping their name on chocolate bars."

Didn't sound like you could make millions of units that way, but I supposed it was still early. The first semiconductor fabs were pretty kludgy, too. Automation kicked in later.

"I guess what I don't understand is what you can do with something that small. I'm playing around in health care and hearing about proteins and tumor cells and all that," I said.

"I've seen some tubes or nanowires that are engineered with binding affinities specific to a certain molecule. You can get a molecule to bind to the wire and change the conductivity of the wire or get a current to pass through it. I think a company named Nanomix does that," Josh said.

I wondered whether every company had the name Nano in front of it.

"Homeland Security is all over this," Josh went on. "They're interested in biosensors for detecting single molecules of bio samples or gases. Pretty neat stuff."

I agreed, but still didn't see how this was going to mess around with health care. No *Fantastic Voyage* ships or self-assembling white blood cell clones or mini-laser zappers. I must be missing something.

Part VII

Chapter 49

∎

Occult

My fecal occult kits finally arrived while I was up in Seattle. It had taken forever. Was there a run on these things?

I won't give you all of the dirty details, but I will say there are a lot more pleasant tasks than throwing a piece of paper into the toidy, next to your business, and waiting around to see if it changes color. . . . I skipped lunch.

Worse, I had a bit of a scare. You do these tests three days in a row. On the second day, I must have waited four or five minutes instead of two and came back. No color, but then I looked close. Not fun. Was that just the faintest blue-green color? It was—it wasn't—it was. I closed my eyes, reached in, pulled the paper out. Very, very faint, but it was blue green, not white.

I sweated for 24 hours.

The next day? Whew. No color at all. When you are done, you test the entire package by dropping some crystals in and the last paper is supposed to turn blue green. Sure enough, it was a deep dark blue and even darker green. It made my faint observation seem pretty stupid.

I was clean, at least as far as fecal occult blood tests are concerned. I was still considering a virtual colonoscopy, but the idea of getting pumped with carbon dioxide for 20 minutes makes the doodie duty seem rather civil.

Most everything I'd read about these fecal occult blood tests suggests they do a pretty good job of warning for colon cancer. I supposed my genetic tests could be wrong. Or maybe I was just a ticking time bomb. It would be nice to know for sure.

Back in 2001, I think, LabCorp, the guys with the walk-in blood test center that wouldn't draw my blood without my doctor's signature, did a deal with a company named EXACT Sciences. Based out of Burlington, North Carolina, EXACT had a DNA test for predisposition to colorectal cancer.

The press release said, "DNA from tumors is shed into the colon and carried out of the body in stool. LabCorp will provide a non-invasive, patient-friendly method for patients to collect whole stool samples in the privacy of their home." Something the whole family can do together? Basically, you crap in a bag and instead of setting fire to it, putting it on your neighbor's doorstep, ringing the bell and running away, you mail it to LabCorp. I bet the mailmen were pleased.

From what I heard, the tests were no better than fecal occult blood test kits, similar to the one I was using, and so yet another interesting genetic testing idea got, er, flushed away. EXACT stock was $15. Last I checked, it was $2.

There is a lesson in that story somewhere, but I'm too grossed out to dig into it any further.

Chapter 50

■

How Much
Are You Worth?

Those fifth-grade Mr. Science movies or *This Is Joe's Body* always had a scene with a nerdy but brilliant scientist holding up a glass of murky liquid and declaring that the human body is really just $1.25 worth of chemicals. Of course, Nixon was in office, and much has changed since then.

But what are you worth, really? Wrongful death suits can run into the multimillions, but that's probably just a quirk of skewed laws, skewed to lawyers, anyway.

The 9/11 Victim Compensation Fund had the task of disbursing $7 billion—$1 billion to those injured and another $6 billion to the families of those who died that day. 2,878 families received payments averaging nearly $2.1 million apiece, tax free. Families of young bond traders received a lot more than average, based on their age and life expectancy and estimated earnings power based on their latest income and bonuses. Families of the 25 wealthiest victims, people who were making a couple of million bucks a year, not a stretch on Wall Street in 2001, were given awards of $6.3 million. I assume this means that a poor 70-year-old janitor's family received $1 million, or less.

That's an interesting range, $1 million to $6 million. In health care, it turns out, the number is not that much different, if you can get anyone to actually admit it.

As Dr. Baker at Stanford explained to me, it's a question of how much a quality life-year is really worth. No one knows, or more likely, no one wants to say. Almost every article you read on the subject puts the range at $50,000 to $100,000 per quality-adjusted year. That's how much Medicare might consider paying for drugs or treatments.

Plug in life expectancy, say 80 years, and you get $4 million to $8 million in the value of your life, at least as Medicare and insurance companies look at you in raw numbers. Is that too little, too much? Hard to say.

That's how insurance works. Their bet is that enough of us are going to die from something else, hit by a bus, natural disaster, homicide, or even a quick, massive coronary. Spending $50,000 or $100,000 per year for each and every 300 million Americans is $15 trillion. So obviously, the assumption is that they'll spend that on only 10% of us. But which 10%?

John Simpson is going to clean out my heart with his Roto-Rooter. Some mouse is going to get fried so I can have the antibody I need. It all goes out of whack as we live longer.

Think about it. As we live longer and die from things besides the Big Three, the formulas break down. Maybe we save the money we no longer spend on heart and cancer and overall spending drops to $1 trillion and Medicare spending goes down and we balance the budget and live happily ever after.

Or does this mean that the same $50K per year is going to be spent, but on 30% of us instead of just 10% of us, and health care spending triples? Ouch. Or is the pie finite, and there is going to be only $15K to $20K available per person per not-so-quality-adjusted year to spend on the same 30% of us? I'd kinda like to know, wouldn't you?

Then there's the case for spending more. In 2005, Kevin Murphy at the University of Chicago won the MacArthur Foundation "genius" award, $500,000 over five years, about the same amount Medicare might spend on him if he were sick.

Working with colleague Robert Topel, Kevin Murphy calculated that reduced mortality from heart disease since 1970 has been worth $1.5 trillion to the U.S.

In other words, reducing deaths from heart disease has basically paid for all of health care in 2005. Well, that's not exactly right. It's not the increase in GDP that counts (as the Chinese will find out), it's profits, but the intent is astounding. Saving lives may be expensive, but you get paid back with increased output.

Murphy and Topel go on to calculate that a 20% drop in cancer deaths would mean $10 trillion in value to the U.S.

How do they come up with these numbers? It's a combination of how much someone might spend to increase their longevity plus their increased consumption and economic output. And wouldn't you know it, Murphy and Topel suggest the value of the average human life might be between $3 million and $7 million, so they use $5 million in their calculations. I'd like to think that Americans are all above average.

The numbers can get huge. Eliminate heart disease altogether? $47 trillion in value. No more cancer? $48 trillion.

If ever there was a case for increased spending on early detection, Murphy and Topel have it. Of course, that's like saying tax cuts increase tax revenue. They probably do, but no one who sets tax rates actually believes it, so the funding battles continue.

Chapter 51

■

Cancer Conspiracy?

"How much do you think is spent on early detection?" I got back in touch with Don Listwin, figuring he would be the one who would know.

"Compared to all cancer research?" Don asked.

"Sure."

"Drop in a bucket," he said.

"10%?" I asked.

Don Listwin started laughing. "I'd be thrilled, and doing something else with my time than running up to Seattle and looking at mice models."

"So how much?"

"We don't really know," Don said. "Probably less than 5% of the NCI budget."

"National Cancer Institute?" I asked.

"Yeah." Don thought for a bit. "But if you add academia and pharma to NCI, it's probably less than 2%."

"That little?"

"A five-year grant for biomarker work of about $20 million on a $4 billion budget, what's that?"

"That's half of one percent, I think." I could still divide 20 by 4,000 in my head.

"There you go" was all Don could say.

• • •

I can understand the cholesterol conspiracy. No one is really hurt if our cholesterol is lowered. Even if it costs $25 billion a year out of our pockets and countless hours fretting over our very existence and survival, what's the big deal? And death rates from heart attack and stroke have been dropping. Well, they've been dropping since the 50s, well before Lipitor, but details, details.

As I found out, death rates from cancer have been, pardon the expression, dead flat since the 50s at 193 per 100,000 people. Billions and billions are dumped into research for cancer care, drugs, chemotherapies, and on and on. Rightfully so, because people are dying today from cancer. Anything we can do to help them is humane and necessary. I'd like to think if I got cancer, or any member of my family did, there is some cure out there.

But early detection gets shortchanged in the process. Why is that?

Is there a cancer conspiracy? Certainly hospitals make tons of money treating cancer patients, an almost open-ended spigot of reimbursements from insurers and government programs. Same with pharmaceutical companies. Each potential cancer drug, which may treat only 20% to 30% *at best* of certain cancers, can generate huge sales and profits. A couple of grand a month in drugs to treat patients, to extend their lives by six months, a year.

Worth it? Of course. But maybe we're at a point where early detection is the right way to go. It becomes cheap enough to screen all of us for cancer and other diseases. Treatment can be painless and dirt cheap. A zap here, a clip or snip there. Voilà, you're clean.

Does everyone in the health care industry want this? I don't see a lot of hands going up volunteering to fund it to make it a reality.

The NIH and NCI? Blood from a rock. Pharma? If it's there, no one can really see it. Hospitals? Nope. Academia? A little, but it's probably a career killer, as Sam Hanash points out. Cancer centers? Starting to.

The Hutch may be doing the most, needing an annoying—in the nicest sense of the term—outsider, Don Listwin, to kick butt and take names and reach into his own wallet to fund this wacky but potentially life-altering direction.

$20 million is a huge dent in early detection funding, but just noise in terms of overall cancer spending. I mean, come on. Microsoft spends $6 billion on research and development, Intel another $5 billion. In 2006, NCI's budget is just north of $6 billion. About half the money is grants for research projects—how much of that is care versus detection isn't clear, but probably 95 to 5 at best. Another billion is research support and management plus another couple hundred million on training and education. In poring through their budgets, I can't even find an early detection category—it gets lumped with prevention and prediction. Huh?

Politics? Vested interests? Who knows, but it doesn't seem right, does it?

Sure, people are dying. It's awful. But is the money being spent efficiently? Are these guys killing us by not spending on early detection? It sure feels like it.

Chapter 52

■

How Big Is an Antibody?

And then it hit me. Maybe like CT imaging and the 3-D models that go along with it, R&D money was already being spent on health care—we just don't count it that way. Intel spends $5 billion a year on R&D and all sorts of billions on factories to crank out their Pentiums and memory and all these chips for PCs and cell phones and everything else. If there was some medical chip, as Dr. Gambhir suggested . . .

I picked up the phone and called Sam Gambhir again. He must have thought I was a stalker, I'd been bothering him so much. "Sam, can I just ask you a quick question?"

"Sure, what's up?" Sam hadn't tired of me.

"Okay, I've had two tours of your imaging lab. I've seen molecular imaging work. Some probe got injected and within five minutes I could see a breast tumor from the MicroPET machine. I assume that ought to work for humans, too?"

"Is that your question?"

"No. I've also gone to the Hutch up in Seattle and seen all the work they're doing in mass spec and finding these proteins and making monoclonal antibodies for biomarkers."

"Yeah, we're doing some work with them. You know Don Listwin?"

"I've just spent lots of time with him," I said.

"He's got us doing some imaging work with the Hutch. Lymph nodes—quite interesting—something important could come out of it."

"So if they find all these biomarkers—" I started.

"We need all those as the first screen, and then we can do the custom probe for imaging."

"I think I understand that. But here's what's confusing me. It's all biology. I'm just banging my head against the wall trying to figure out how this stuff scales. There's got to be some silicon somewhere?"

"And?" Sam asked. I was hoping he would just spill the beans about his grant.

"So I'm missing some data. My question is, how big is an antibody?"

"That's it? That's easy. It's about 30 kilodaltons," Sam answered.

"My scale doesn't have Daltons."

"Its atomic mass? Oh, a Dalton is basically one-twelfth the weight of a carbon atom. One over Avogadro's number grams?" Sam was testing my high school chemistry.

"Grams? No, no. Not its weight, its size. How large is an antibody?"

"It's all related. Let me put it this way. It's several thousand amino acids."

"Still not getting it," I pleaded.

"An antibody is probably about 20 to 30 nanometers in diameter," Sam told me.

Eureka number five! I still wasn't getting used to these moments. But there it was: you don't need room at the bottom—you don't have to get much smaller than 20 nanometers—you just have to somehow interface with antibodies at their own level.

"Got it, that's a number I can use. Now it all makes sense. That's the key, isn't it?"

"We are just about there," Sam said. "Come on in when you're ready."

When I started following the tech business, back in the early to mid 1980s, chips had dimensions of around three to five microns, or millionths of a meter. In 2005, the first set of chips with 65-nanometer dimensions started rolling out of fabs. That's about a 50-fold decrease in dimensions over 20 plus years. One more halving and we are at about the size of an antibody. This is what George Gilder meant. When you get close enough to the size of interesting molecules, you no longer have to shrink dimensions—you just have to start connecting these to real-world things.

I've been a skeptic of nanotechnology—which has been overbilled as self-replicating nanobots that could be used to build space elevators.

But this may be it. Intel is already showing 45 nanometers in the lab. At 30-nanometer electronics, you can start querying antibodies. I'm not even sure what that means, but I think I'm on the right track.

Chapter 53

■

Sam Again

I was back in Sam Gambhir's office. I knew I was close. Maybe he was ready to talk about this chip of his. I'd figured out he could connect to antibodies—but wasn't sure how or why. Was it for imaging? Probes? How did this work?

"I think I get the concept of molecular imaging, but I'm still scratching my head on how it scales," I started.

"Maybe in a lot of ways," Sam asked.

"I'm trying to look at it from the cost side. Does this get cheap? I mean, I can see with CT scanning how we go to more detector rows and 3-D models and the whole thing scales. But with functional imaging, cancer is either there or it isn't there," I said.

"Right," Sam said.

"So," I said, "it seems like great stuff. But I haven't heard the Silicon Valley numbers—4, 16, 64, 256—like I've heard with CT scanners."

"Right," Sam said.

"So if it costs too much, insurers and Medicare will just pass, won't they?" I asked.

"Well." Sam thought for a bit. "Actually, it is cheaper than the alternative."

"Such as?" I asked.

"Well, treating cancer isn't cheap. Early intervention ought to be cheaper."

"So you think there's someone out there with a giant spreadsheet that says it costs $50,000 to treat a cancer patient, so they'll pay for early detection and molecular imaging only if it's $500 or $50?"

"Probably. That's just reality," Sam sighed.

I was coming to that conclusion, too. I'd found some really earth-shattering medical technology and it wouldn't be deployed because of cost? Heartbreaking.

"So it could—" I started.

"Of course, there is a way for this to get really cheap," Sam said excitedly.

"How so?" I asked skeptically.

"If we can do some sort of screening up front, then we can know what type of tumor is present. From that, we can formulate a specific probe to find it and maybe even a drug that attacks just the tumor."

"Like the copper-64 stuff, upping the radiation in the probe."

"Maybe. But more likely a very specific drug. We can work with pharma to show exactly where their drugs go and help them target them a lot more specifically, especially if they are hanging off probes. If we can pull this off, dosages go way down. It would be like licking an aspirin and throwing the rest away. That's all you'd need."

"But what do you mean by some sort of screening? How can you figure out what to target?" I asked.

"Did you know about U-54?" Sam asked.

"German submarines?" I asked, much too quickly.

"No. No. It's our grant application to NIH. NCI, actually. Did I tell you about this yet?"

"A grant for what?" I asked.

"For our antibody chip."

"So this is the chip I keep hearing about?" I asked.

"Could be." Sam had a little smile on his face. "We just got it approved. $20 million over four years."

Hey, some money for early detection.

"We got it approved by linking it to drug discovery. The grant is for a chip to screen up front and then for molecular imaging to determine whether drugs are effective."

"Because?" I asked, wondering why there had to be a drug discovery angle.

"Because that's how you get grants approved."

"I get it." Sadly, I was starting to figure out how this funding game works. If you don't bill it as early detection, it has a fighting chance of getting funded. "Okay, but is this like the Affymetrix GeneChip?" I asked.

"Theirs is optical. We need to do it in silicon, to make it really cheap to make and use. Ours is not for labs but for real people."

Sam got up and walked over to the other side of his office. He picked up a four-inch-thick notebook stuffed with papers ready to bust out all over the floor if he dropped it.

"Have you ever seen a research grant?" Sam asked.

"Not really," I said, wondering how long it takes to generate four inches worth of paper just to request money.

He opened it up about two thirds of the way through.

"See that? That's our matrix." Sam pointed to an illustration in the grant request—it was of a dozen tubes with holes in them, in a crossing pattern, with small blobs at each intersection where tubes crossed.

"Think of this as a matrix of antibodies," Sam said.

"Okay."

"There are thousands of potential proteins as biomarkers, right? We want to use them all. Whenever an antibody and an antigen combine, they give off an electrical charge."

"They do?"

"Sure, a tiny one, but we can measure that easily. As blood or urine or whatever flows through, the antibodies will identify what proteins are in the serum, and depending on the point in the matrix where the charge appears, we will know exactly what protein was present."

"But how do you do that?" This was amazing.

"We attached the antibodies to tiny wires arranged as a matrix."

What was the matrix? I was closer.

"So?" I asked.

"Remember what I told you. Antibodies are 20 to 30 nanometers in size. We can make connections in that size neighborhood. And the technology exists to attach different antibodies to each of the intersections."

"But how?"

"We've got Hong Dai and Jim Heath and Bob Sinclair involved in this. You know these guys?"

"Wasn't Jim Heath involved in the buckyball thing?"

"That's right. All of these guys studied under Richard Smalley at Rice. Heath in particular has focused on molecular switches—nanowires connected to molecules. It's not that different from the way Affymetrix puts DNA strands on their glass, but we use silicon. We attach to electrical conductors that can carry current when proteins and antibodies connect."

Sam paused and took a breath.

"The technology exists—it's just a question of building it and getting it to work."

So it's nano. Makes sense: 30-nanometer-size antibodies connected to a 30- or 50-nanometer nanowire or nanotube. Brilliant.

"So it's a giant protein sensor?" I asked.

"Sensor, yes. Giant, no. Think nanosensors. And the chip won't be that big. Even if we have 10,000 antibodies, it still won't take up that much room. This chip could be really cheap."

"Like a nickel?"

"In volume, aren't all chips a nickel?" Sam asked.

"Yup," I said. "It's odd that Smalley had cancer—in a weird circular way, his invention may be the key to early detection of cancer."

"Hmm," Sam hmmed.

I sat for a moment trying to absorb the impact of all this.

So far, it's all a bunch of chemistry, mass specs to weight proteins, and find biomarkers. I had assumed that like the tests of today, it would be the ELISAs that Don Listwin had told me about: $10, $20 each. And then an ELISA for each different protein. No way was that going to scale, but at least it was an early warning system.

This changed everything. If Sam could really attach antibodies to tiny wires . . . holy shit, this is what George Gilder was talking about! You can't theorize it—you just have to build it.

"So this is the ELISA killer, isn't it? Those tests become old-fashioned?"

"Sort of. ELISAs are too expensive, anyway. You can't expect people to pay $10 each to test for hundreds of different proteins."

"But thousands at the same time? Will the numbers work?" I asked.

"Well, think about it. If you do mass screening up front, everybody is tested with these cheap chips. Every year, every day, whatever. More so as it gets cheaper. We'll know exactly which protein is present. We can almost make custom probes to go in and image tumors and custom drugs to kill them. Almost by definition, treatment is cheaper. This is true personalized medicine, not the dreams of genetic engineering."

"So imaging plays a role in a larger process?" I asked.

"We'll probably always have some form of imaging. Doctors like to see pictures. The trick is working with pharmaceutical companies to come up with the right probes. These pharma guys may not like that we find all of these cancers so early, but you can bet they'll have an interesting business developing tracers and personalized probes and custom drugs. It's quite different from what they do now. Plus, the diagnostics business could be huge. First, there will always be new cancers and new proteins and antibodies to use with the chips. Plus, there are 6 billion people out there. Put a chip in every cell phone. Or at home, you could have these chips in toilets . . ."

Flush to crush cancer.

". . . or someday we might even be able to implant them."

"That's a bit out there," I protested.

"Perhaps. But remember, there is no *Star Trek* Tricorder. This is actually better," Sam declared.

Chapter 54

■

Cheap Chip

Sometimes life really is circular.

I could have guessed this a long time ago and saved all the hassle and aggravation. I'd been to hospitals, followed doctors around, got poked, prodded and scanned, watched mice twitch away in boxes, sat through 3-D imaging face-offs, visited cancer centers and state-of-the-art CT clinics, all in the name of finding something in medicine that scales like Silicon Valley.

So what did I find myself doing? Driving into the parking lot at 2200 Mission College in Santa Clara, California.

I'd been to this place a zillion times. I spent way too many years on Wall Street following this company, going to meetings, getting tours of their fabs that included crawling around underneath them. I'd had just about every vacation interrupted by their stock blowing up, and reveled in the success of having a right long-term view as their stock soared. But that roller coaster was over for me. I laughed now when they glitched, knowing dozens of analysts are having a bad hair day. Not me. I never wanted to think about them again. I was done. Which is a big reason I was sniffing around in proteins and biomarkers and imaging in the first place.

I shook my head and could only groan as I passed a sign that read INTEL.

I was here to meet with Andy Berlin in a group known as Precision Biology. A little bit different than microprocessors.

"The tools that Intel uses in making chips are so finely tuned to tiny structures that they have uses in biology and health care. One of the first things I discovered when I got to Intel was that they actually have 200 analytical chemists at the company. One way to look at this place is as a giant chemical processing factory, lots of expertise here."

"And you try to figure out how to apply that to biology?" I was trying to get Andy Berlin to tell me more about chips and nanotubes and antibodies, but maybe there was more to it.

"I think it's more than that. We have a laser imaging system that scans chips to find defects—tiny imperfections—in them. It turns out that the Raman laser in this system subtly interacts with molecules, enough that we can make a decent spectrometer to get fingerprints not only of proteins but of modifications of those proteins in blood serum."

"How they change?" I asked.

"Communication in our bodies is via modifications of proteins—attaching sugar to them, for example, might signal cells to replicate—and other ways."

"And the laser—" I started to ask.

"We can see those modifications. Nothing else exists today to see this."

"Faster than a mass spectrometer?" I wondered.

"Probably faster, but the things we can detect get stripped off—for example, the sugar falls off—on the way into a mass spectrometer. We have a lab up at the Fred Hutchinson—"

"The Hutch?" I asked.

"The Hutch in Seattle," Andy said.

"I've seen their mass spec labs. Those things are slow."

"But that's for researchers. There's probably several stages of deployment. Research labs, doctors' offices and then homes."

"You see a day with a chip cheap enough for homes?"

"Sure, the chip is just part of it. Think about something that fits into the palm of your hand, plugs into a PC, the sophistication of a CD burner, with similar costs."

"As the CD burner? Those are pretty cheap," I said.

"$30 is reasonable."

"So it can happen?" I marveled.

"Depends on what you mean," Andy said.

"Sam Gambhir at Stanford is talking about an array of nanotubes and antibodies bound to the tubes and—"

"I'm familiar with Sam's work. His array is quite clever. But you should know that even a chip with nanotubes and antibodies is only one small piece of a bigger solution in health care."

"But it's the piece that scales—gets cheaper over time, right?" I asked.

"Probably, but the whole thing needs to come into being. There's four main things to focus on. First, we need a feedback loop of treatment correlated back to medical records. Patients who tested positive for this biomarker and took these drugs had these outcomes. The data is there, but it's not in a form that we can mine it and get what we want out of it. If you think about it, Intel has perfected a huge failure analysis model for making chips. We characterize every step of the process and study things that go wrong and make changes. There is a huge database to mine. Health care can be a failure analysis system on people. It's what's missing. We think we know how to do it. It's an IT problem."

Hmm. Maybe electronic medical records are for more

than just saving money. Still, we need records online before you can even think of mining them for juicy data.

"And?" I prompted.

"Next, we need better nanotechnology to look at serum. And at drugs, actually. We can attach things to molecules and then get them to light themselves up. A bit like the molecular imaging Dr. Gambhir is doing, but with some new techniques. Think of this step as instrumentation. Biomarkers are just one quarter of the solution. There is a discovery process that needs to be industrialized. We've putting labs into another three to five research centers over the next year or so."

"But that's just instruments, labs, a process," I said.

"That's right. Then you need the content. People asking the right questions. What biomarkers work, what drugs work best, what's in our blood that we can measure—someone using increasingly sophisticated instruments to generate more markers and more pharmaceuticals that can have measurable results."

He paused and took a breath.

"And that leaves?" I asked.

"Deployment."

"Some piece of silicon. Antibody binding?" I asked.

"We'll see. There are lots of ways of doing this. Binding lots of antibodies may create some issues. Did you know there's antibody cross-talk?"

"No."

"Some antibodies stick to other antibodies' proteins. They mess with each other's signals. It gets complicated."

"So an array of thousands of antibodies?" I asked.

"Lots to be worked out. Stanford has a measurement system that can determine which protein is bound to which antibody. There's also similar technology now used in optics—MEMS, tiny mirrors—that can be adapted."

"You have to build it and see. But isn't that like the early days of Silicon Valley? Try something, perfect it. Open the door of the oven to see if it's done."

The more I looked, the more it felt like those early days to me. Something was possible, there were issues, and capital funded a bunch of different approaches and experiments raged through the night and innovation exploded all over the place. Maybe it wasn't Intel that would provide the solution. Maybe, as Don Listwin predicted, we'd see venture capitalists coming out of the woodwork once a return was visible. Thousands of biochemists descending on Silicon Valley, or anywhere for that matter, finding biomarkers, developing protein sensors, imaging probes, outcomes databases, all the little pieces that are needed. No one company provides everything in PCs or cell phones or certainly the Internet. It's a big stack of turtles.

"Could be. Silicon Valley is often about turnaround time between experiments. Remember, biology is slow. You have to wait for lab results. Patients get well slowly. It takes time," Andy Berlin reminded me.

Well, there was a dash of cold water. It felt like Andy Berlin was talking me down from high expectations. "So I'm too optimistic?" I asked.

"No way. I have every confidence that this will happen. It's a brand-new field. But it's not about some new chip, it's about a system of how to convert health care. People in medicine don't like it when I say failure analysis model, but that's really it. You'll have that $30 nanoscale electrical-ELISA wrapped in plastic. And the system it talks to. I think it's a realistic five to ten years for this to play out."

Sam Gambhir and Intel weren't the only ones trying to detect proteins with chips. I connected with Dr. Charles Lieber, a Department of Chemistry and Department of Engineering professor at Harvard, who is doing similar work.

"We already use nanowire arrays for protein detection. My group has pioneered and demonstrated that they can be directly applied to blood scrum samples with great sensitivity and selectivity. In a work soon to be published, this has been extended to much larger arrays. It is really a very well developed science and I don't think anything else needs to be invented. As an aside, we've also applied this to DNA detection, small molecule/drug screening and virus detection at the level of single viruses."

This was it, and probably more. The chips that everyone kept referring to were really part of a bigger system. Bleed to read. A sensor for cancer. But perhaps connected via your PC and cable modem to a system that constantly mines patient databases for outcomes and suggests proper drug therapy, or if early enough, the zapping machine.

If you had told me this even six months ago, I would have laughed it off as a Buck Rogers, pie-in-the-sky future. Isn't that always the way it is with technology?

All the pieces aren't yet in place, and Andy Berlin's five to ten years may turn into 20, but . . . early detection just got that much closer.

Chapter 55

■

Virtcol

Okay, I'm ready for Sam's antibody chip to become a reality. Right now. Since failing that stupid genetic test for colon cancer, I've been putting off and putting off a real test. I blamed it on needing to do more research on a real colonoscopy versus a virtual one. I decided on a virtual colonscopy, to have something to brag about at cocktail parties. Turns out, the test itself is nothing. It's the night before.

"Okay, here's a prescription for Colyte. Follow the instructions carefully and then don't eat anything afterward until your test is complete."

"That's it?" I asked.

"Yeah, uh-huh, pretty easy," the nurse told me. It wasn't hard to detect her dripping sarcasm. I had no idea.

I paid for the Colyte—it said Golytely on the package, close enough, I suppose. There was no skull and crossbones, but it probably should have warned about explosions.

I filled a gallon jug with water that dissolved the powdered mix inside. The instructions said I should drink a glass every 10 minutes, which would be followed by eliminations. I never finished the gallon. I did, however, experience a complete evacuation of my facilities. I did my best impersonations of Rodin's *Thinker*.

• • •

"Have you done your eliminations?" the nurse asked me when I arrived for my procedure.

"Oh, yes."

"Good. Now lie down on the table, facedown, please. We will be introducing carbon dioxide to inflate your colon."

I won't describe in detail what happened next except to say I couldn't resist mumbling, "Shouldn't we get to know each other first," under my breath.

"Okay, now hold your breath, this will only take about 20 seconds."

Sure enough, I moved through the ring before I even had a chance to think about it.

"Turn over, please, we need to do it faceup as well."

I turned, held my breath, and like that, it was over. Despite an empty stomach and colon, my appetite had disappeared.

Again, the waiting. This time, like Rodin's sculpture, I did have a lot of thinking to do. It's not the end of the world to have colon cancer. It's easily treatable. Unfortunately, that means having a real colonoscopy, and repeating all of those wonderful steps.

But then I remembered why they do colonoscopies. Because they can. Maybe I have colon cancer and maybe I don't, but it is pretty easy to tell because you can actually go in that den of destruction and look around. No such luxury in my pancreas or liver or kidney or prostate or top of my spine or brain. Doctors can look, but probably won't see a tumor unless it is already big enough to kill me. What's the use of that?

"Mr. Kessler?"

"Yes?"

Again the stone face. No emotion. The doctor greeted me and led me to his office. I could be a dead man walking.

"Let's have a look. Fortunately, there is 3-D software that

enables me to fly through your colon." I nodded my head. I had already peeked into someone else's colon. It was a little more personal with your own.

"First, I want to congratulate you on a wonderful job with elimination. It is quite clean in there." Was that really necessary? Jeez. "And you'll be happy to know that you are as clean as a whistle." Whew. It felt like a 100-pound sack was just lifted off my back. "No polyps, no lesions. You look pretty good."

"Thanks, doc."

But was I clean? Who knows. Maybe the tumor is just tiny, and not detectable. Should I go through this again in five years? Next year? I don't think I could handle the emergency evacuations too many more times.

It was a mixed blessing. I'd had a heart scan and a colon scan and both are negative. I'm clean. Great. But all this time watching mice twitch and proteins smashed in mass specs and studying 3-D imaging and nanotubes was a bit unsettling. And what about the rest of me? I'd like the future to happen a lot faster. No one can give you a clean bill of health unless they test for everything. But, at the end of the day, that's what I wanted. That's what I demand.

Let's get going. Anyone obstructing this goal needs to chug a gallon of Colyte, have a seat and get the hell out of the way.

Chapter 56

■

Critical Path

T he FDA is not known as the hotbed of innovation."
There was laughter throughout the room.

"There is a gap between science and patients who are sick. Molecular imaging is incredibly exciting—wonderful possibilities for medicine. The problem I see is the barriers to the transition of the promise of imaging to real treatment and real diagnosis for real patients in Kansas."

I don't think Dr. Gary Glazer could have said the words any better. But he wasn't the one saying them. Instead, he had introduced the speaker, a woman by the name of Janet Woodcock, who was the deputy commissioner for operations for the Food and Drug Administration.

I was told beforehand that since there hadn't been a sitting commissioner who has lasted for more than 12 or 18 months during the entire Bush 43 administration, Janet Woodcock was running the show.

I had snuck into the back of the small conference room at the Lucas Imaging Center. The discussion had just started. Janet Woodcock and another guy from the FDA, George Mills, had been given a tour of the facilities and the promise of molecular imaging. It sounded like they were blown away. I wonder if they were treated to twitching mice or better yet, made to look into that mouse slicer machine. Probably not on the tour.

"We know there's a bottleneck in pharma drug development. Despite all sorts of new technology in drug discovery, the number of applications for new drugs has gone steadily down over the last 10 years," Janet complained.

Dr. Glazer jumped in, "It used to be that drug companies would do research on 5,000 compounds and none of them would turn into useful drugs. But now it's been said that because of new techniques and more efficiencies in their system, they can do research on 50,000 compounds and turn out no useful drugs."

The laughter rose again in the room.

I noticed from the program that someone from Pfizer and someone else from GE Imaging were participating. On the other side of the room was Dr. Sam Gambhir. I knew he had an opinion worth hearing—this would be interesting.

Janet said, "From what I've seen today, we could select patients with high probability of successful treatment, we could find responders versus nonresponders, we could adjust dosage for specific patients, even test multiple drugs and measure their effectiveness," One by one, she looked at everyone in the room. Must be part of the Beltway ritual of gaining consensus. "But we don't. It's all empirical, all anecdotal."

A lot of heads were nodding. Janet's eyeball to eyeball seemed to be working. I'd have to try that. Only the guy from Pfizer was shaking his head no.

"It's crystal clear to me that empirical testing and the cost of serial, randomized trials—well, it's staggering. Yet society requires it."

The nods stopped. No one was sure whether to agree or disagree.

"We've gotten resistance from the treatment community

to dosage adjustment by patient." Now all the heads, except for Pfizer's, were shaking in disapproval.

"I've put in place something called Critical Path in the FDA. We know there's not a strong business model for biomarker development compared to drugs, anyway. We know there's no rigorous pursuit of qualifying markers—but biomarkers must be used. So we are encouraging a simplified approval process. We're encouraging drug development companies to file their markers when they file for drug approval."

I wondered what "encouraging" really meant. And why all the talk about the treatment community and drug development? I thought the focus of all this was early detection. Maybe this was the cancer conspiracy. It's as if she were talking to a room full of lobbyists for drug companies rather than researchers taking a break from knocking down mice. I glanced over at Sam Gambhir and he seemed uncomfortable.

"From what I've seen on this tour, visual evidence is probably the best thing we have going versus some proteomic data that correlates loosely to some outcome. Show me that and Critical Path becomes a success," Janet finished. Dr. Gambhir sat up.

Okay, now I got it. Show her a breast tumor found using RGD peptides and a PET scan and she'd make sure it got noticed and try to ramrod it through approvals. A picture's worth a thousand proteins.

Her deputy, George Mills—can deputies have deputies?— went through some of the details.

"Many biomarkers are invented just for clinical trials and then disappear when therapy fails or the trial closes. We want those. Imaging is the key to drug development going forward. They are the ultimate surrogate—rather than survival rates or asking patients how they feel." Jeez, is that all they really do. Some empirical evidence.

The guy from Pfizer asked a question about Phase II trials

and the specifics of the wording of the applications for Critical
Path.

Gary Glazer interrupted the discussion. It was his turf
at Stanford and I think he wanted to drive the discussion, so
he jumped in with "It seems like what Janet is saying is that
we have a more permissive environment for looking at things.
So why don't you use more markers and imaging in your
trials?"

I think Pfizer-man figured out he was in enemy territory.
"We'd love to use them in early development, in dose selection,
in efficacy and safety profiling. But at the end of the day, the
FDA wants to know how many bodies we had in our trial, not
some surrogate end point outcomes."

Janet Woodcock would have none of it. "My beef is different. Journals, especially post-Vioxx, are thundering for
more patients, higher body counts. But society can't do this, we
can't afford it anymore. Well under 10% of cancer compounds
that enter Phase I make it to market. And worse, we put drugs
on the market that we don't know much about. 12% efficacy is
no longer acceptable. Better science and this technology is the
way forward." As opposed to spaghetti against the wall, I suppose.

But why can't the FDA just bully this through? It seems as
if the inmates are running the monkey house.

Dr. Gambhir brought up a touchy subject. "Do I have to
go to Korea or China or India to do human testing? It seems
like a moral dilemma if we can develop probes here but can't
test them, and need to go overseas to laxer regulatory environ-
ments."

George Mills shot him down pretty quickly. "We wouldn't
have put in this exploratory Critical Path if I didn't want
human testing in the U.S. I just want as low an on-ramp as possible. Look, you wouldn't believe some of the applications we
get—12 patients tested in Belgium who say they feel better—

and they want approval. If you can show the science and imaging and safety, I'll work with you."

The discussion ended and everyone split up. I saw Gary Glazer and Sam Gambhir chatting afterward. I would have liked to be a fly on the wall for that conversation, until I realized that I was a human on the floor, so I sauntered over and listened.

"What did that guy from Pfizer say they spend on R&D, $6 billion a year?" Gary Glazer asked.

"Something like that." Sam Gambhir nodded.

"About the same as NCI," I added.

"For coming up with almost nothing. We can change that." Gary shook his head.

"$100 million would be a drop in the bucket for them." Sam sighed. I got a sense he would be happy to take $100 million off their hands. And probably do something pretty interesting with it.

"So why is it all about drugs and pharma? It seems like they more or less told you that you have to piggyback on drug approval," I asked.

"That's the way it is with the FDA. That's what they do. They keep insisting that our markers and probes should be treated like drugs. But they're not drugs—not even close. We put in such trace amounts—there's nothing toxic. We can't go through a decade and $1 billion testing each one, even if it's piggybacked with some drug trial. Something has got to change." Sam sighed again. I wasn't sure I blamed him. It felt like the brakes were locking up and early detection and molecular imaging were sprawled across an active runway and big 747s laden with expensive drugs were landing every two minutes.

I hung around for a little longer and noticed George Mills was still there. I walked over and calmly waited for a few others to finish asking questions, and then I got George Mills alone and introduced myself.

"I'm still trying to figure out the business model of all this," I complained to him.

"MBA. '81. Pepperdine," he said.

"Great basketball teams back then," I said.

"Sure were. Janet brought me in thirteen years ago to do this. I was a doc, doing PET and imaging work. I did some clinical trial work on the side. I know imaging is the way to go," George told me.

"But why all the talk of drug discovery and pharma?"

"Welcome to the FDA." He lowered his voice. "Pharma pervades everything. They're everywhere." He looked over each shoulder, either for the effect on me or general paranoia. Or maybe it was just a reflex action? Weird.

"So molecular imaging has to piggyback on drugs, showing efficacy and visual changes and the like?" I asked.

"Sure, a surrogate. I've written the application wording to be, well, palatable to all interests." He looked around again.

I had a flashback. Not a eureka moment, I had a feeling I was done with those. Don Listwin had equated the FDA to the FCC and Big Pharma to regional telephone companies. I thought I now knew what he meant. The regulated control the regulators, via lobbying pressure, congressional oversight, whatever. It was creepy. Around 1990, I visited the FCC and met with a guy in charge of emerging technology. He politely explained how he had set aside several bands of frequencies, wireless bandwidth, for hobbyists and others who didn't have the wherewithal or resources to file for licenses. These so-called junk bands were used by cordless phones and microwave ovens, and he made sure that anyone could use them for any purpose, as long as they didn't interfere with others in the band. And that's exactly what Wi-Fi and a host of other wireless data products are using today, threatening to whack the established telcos and cellular phone companies and maybe even cable. He opened a hole, a back door, for someone to sneak through.

"So, like in basketball, more Princeton than Pepperdine, you're opening—"

"A back door, that's right," George interrupted me. His voice got even softer. "Drug companies complain there's no obvious business model to all this. But the way I look at it, oncology drugs with 10% to 12% efficacy rates—I mean, that's almost criminal, isn't it? Imaging can be the gatekeeper, even give them the model they're looking for. If they can show 30% to 35% efficacy for a smaller set of patients, payers will be a lot more willing to pay and probably expand the market."

Again, all the talk of pharma. "And early detection? Isn't that eventually what this is all about? Slip in your back door and then completely change medicine?" I asked.

George looked to the left and to the right and then looked me in the eye and nodded. He didn't say another word.

Chapter 57

■

What Is Health?

I had a bad flashback, all the way to 1995 (not all flashbacks make it to the 60s). I was sitting in a tall building in Hong Kong talking to investor William Kaye. He was explaining to me about how easy it was to invest in China. The industrialization that was taking place was no different than in the U.S. after World War II. He had seen this movie before. The plot was the same, only the actors had changed.

All those eureka moments woke my sleepy butt up. Seeing digital imaging and plans for a cancer detection chip and a back door open at the FDA for all of this was all I needed. Change was a-comin'. The geeks weren't just at the gate, they'd laid siege and were halfway up the ladders with Treos drawn, ready to attack.

I'd seen this movie before, too.

In fact, I'd seen it so many times in reruns, I had the plot memorized—it was no different from boy meets girl, boy gets girl, boy loses girl, boy gets girl at the end.

Here's the plot.

Things are going along fine. New technology comes along that does things cheaper and better. Barriers are put up to keep that technology out. Technology, instead of going through that nice comfortable way of doing things, and knowing no better, goes around it. On both sides. Like a snake curling up around its prey. When it gets to the other side, technology squeezes, kills the old way of doing things in an industry, and then moves on to its next target.

You know these movies, too. ATMs came along, and all of a sudden, most tellers were out of a job. Oh, sure, they put up a good fight. Banking was too complicated for the average Joe. The public needed tellers or a financial crisis would surely happen. Checks wouldn't get cashed. Little Joey would starve. Uh-huh. But the technology went around tellers and squeezed them out. Meanwhile, it was the best thing to happen to banking, if not tellers. Costs came out, the industry consolidated, lots of banking went online. Hoo-ray.

Do you call a travel agent to book a flight anymore? Not unless you are stupid and want to waste $50. But travel is complicated, right? The public needs travel agents. Not anymore. Has this been good for the travel industry? Airlines? Maybe not, although costs did drop. But hotels and restaurants and double decker bus companies rejoice.

Ask Wall Street traders what happened to them when online markets took hold. Commissions dropped and lots of traders got squeezed out of the business. Bad for traders, probably not great for Wall Street firms, but pretty good for my online self-directed IRA, another piece of technology that snuck around brokers and pension fund managers and a whole host of middle managers who used to manage our savings. Bye-bye to you and the rest of us win.

Same plot. Get the people out. People are the expense.

Magazine layout, retailers, librarians, postal workers, car mechanics—get 'em out. The market will retrain them for more productive jobs.

Technology is liberalizing and democratizing every industry you can think of. The New York Stock Exchange is toast. Your phone company is shrinking. Broadcast TV and newspapers are starving for growth. We can all take it in our own hands and use the Web to trade stocks, make phone calls, read news, even watch video. But why can't we take charge of our own health care?

The old way is getting stale: get sick, read up, hector your doctor to do something.

There are no best practices, just best doctors. It's not an industry, it's a collection of studious folks who memorized the organic chemistry textbook. They're isolated practitioners. Medicine is not vertically integrated or horizontally integrated—*it's not integrated at all!*

Medical knowledge is scattered to the wind—little bits of it in lots of individuals. There is no product—you and I are the product. Medicine consumes us.

But if I learned one thing in the last year, it's that change was starting to take place in medicine, and was only going to accelerate. Laparoscopic gallbladder surgery is cheaper and became the norm. Why? Because it reduced hospital stays. Hospitals are where people work and people are expensive. Doctors, nurses, orderlies, bedpan emptiers, pathologists, administrators—they've all priced themselves out of existence.

Like a boa constrictor, technology is quietly sneaking around these folks and preparing to squeeze the life out of them. They'll be twitching like the mice in Sam Gambhir's lab.

It's pretty simple. They are the reason for high costs. Forget the God complex thing. I'll forgive doctors for that. The real crime is that they don't really know anything. Doctors use ancient tools, memorize symptoms and solutions, and a halfway decent search engine can leave them in the dust.

But now, the days of doctors are over.

How do I know?

Go back to tellers. Almost everything a teller knew was embedded into the software of an ATM. Instead of paying

$25,000 for a pleasant teller who would smile while counting out your twenties, you could put four of them outside the bank and run them 24 hours a day, Sundays, too. Working banker's hours is an old joke.

A website can embed travel agents' knowledge into a series of clicks. Call centers can cost $1 a minute—a travel site is cheaper and better. A diagnostic chip in a car embeds a mechanic's knowledge of what might go wrong.

The plot thickens—while technology squeezes the life out of them, it steals their knowledge and then spits them out.

And that's what's about to take place in medicine. Embedding the knowledge of doctors into software and silicon. It productizes medicine for consumption. Microsoft and Intel did this for computers. The high priests of mainframe computing gave way to empowered users. Their programming expertise became embedded into a consumable spreadsheet product—customize it for your own use. Eventually, the COBOL programmers disappeared, opening up massive new markets.

This is what personalized medicine is. It's about personal medical knowledge. Open you up, take a look, equate conditions to possible outcomes and suggest corrective action.

> "Your scan shows plaque. You might want to have it removed. Call 1-800-ROTO-ROOTER." "You've got this protein in your blood, you probably have cancer. Let's find it and zap it."

Okay, I can hear you saying, "But people can't be trusted with their own health care. It's too complex."

Oh, really?!?!?

It's only complex if you get sick. It's not all that complex to test to see if you are well. And test early enough that the cure is simple, not complex.

Plus, never underestimate the resilience of a focused mind.

Lots of people without engineering degrees figured out how to use PCs, add sound cards, configure DMAs and IRQs, add CD-ROM drives, install software that took days to complete. Now, grandmothers configure e-mail themselves. No sweat.

America Online billed itself as a walled garden, a *Love Boat* cruise ship designed to keep its users away from the shark-infested waters of the Internet. What a joke. It took only five years and all those teenage girls have figured out how to use Yahoo! and Google and Myspace and Flickr and Facebook, all by their lonesome selves. It happens every time, why should getting scanned or bleeding into a chip be all that different? It won't be.

You can smell it from this far away. Doctors are toast. It's the magic pill—heart attacks, stroke and cancer are cured. Except there is no magic pill, but we get the equivalent of it. If health care budges even slightly from chronic care to early detection, the waves of change rip through medicine like CAT 5 hurricanes. Every assumption about how many doctors and what type and what they do comes into question.

I fired my doctor, but what a huge hassle it was even to get my blood drawn. But it's just starting. Radiologists are being replaced by computer-aided detection. Ophthalmologists are being shut out by LASIK. Pharmacists dispensing statins will be replaced by cheaper-in-the-long-run plaque-removing procedures. Dieticians will be replaced by minimally invasive stomach-stapling surgery. Physical-peddling physicians will be replaced by 256-slice scanning machines. And then, we can only hope, cancer specialists get replaced by an antibody-laden cancer detection chip.

Will we see doctors on soup lines or selling pencils out of tin cups on street corners? Are you kidding me? There is so much work to do. When the PC kicked mainframe booty, tech employment went up by a factor of 1,000, maybe more. But it shows up in a different form—in the snake.

It's all about research and development and specialists. Which antibodies do we put on the chip? What patterns are we looking for in the arteries leading to the left ventricle? Which drugs, as proved by imaging, actually kill tumors?

Those who figure this out, the knowledge creators, don't sell their knowledge in eight- or ten-minute chunks. It's not a game of one-off, but of selling in mass quantities. They're the Embedders and the Productizers—the Scalers.

This really could flip. Medicine really could go from chronic care to detection. Just a change in how dollars are spent means we would get a huge growth area in the economy instead of a huge financial burden. A double whammy. The stock market will help allocate capital to this business, rather than some socialist system of sphincter pricers at Medicare in Washington, D.C.

Investors will swarm like killer bees. Once there's a product to consume, one that gets cheaper and cheaper every year, well, then, change is almost automatic. Control shifts. Industries die and new ones are born. Boil some water. Get clean sheets. Call the midwife. Start picking names. This will be the first really new industry in the twenty-first century.

Time to start another hedge fund?

Chapter 58

∎

The New Medicine

So what is this all coming to? Am I even qualified to take a stab at this, as an outsider looking in?

Maybe so. Sometimes it's hard to see the sclerosis when you're a doctor moving through the system. Plus, my arteries and my colon are clean. Perhaps that makes me objective—more so than if I was dying of cancer and demanding bottomless spending.

In an era of readily available and affordable scanning of our arteries, Stanford's head of radiology, Dr. Gary Glazer, came out and declared to me that heart attacks and strokes are a thing of the past. Dr. John Simpson, who perfected angioplasty, which saves millions of lives every year, isn't satisfied, and insists that if he knows about plaque, he can get it out. Do we scratch heart attack and stroke off the list of killers, much like we did smallpox and polio?

The FDA has snuck open a back door for Dr. Lee Hartwell and Dr. Sam Gambhir to slip in early detection and molecular imaging that may soon—5 years? 10 years? 20 years?—take cancer off that killer list as well.

I've got to admit, I set out to see if I could find some piece of technology that might help medicine get cheaper and scale like Silicon Valley. But I discovered something much bigger.

Never in my wildest dreams did I think I'd trip across a Ponce de Leonesque fountain of living.

Because that's what it's all about. It's not a Fountain of Youth—it's not about staying young—it's about staying healthy. They say 60 is the new 50. If you stay healthy, got a good ticker, lay off tobacco, are lucky enough to avoid some weird cancer, you can kick up your heels, keep running your company, or better yet, travel the world, hike a mountain, ski Zermatt—heck, Tony Randall even started a new family.

But that's a big if. We pump ourselves with cholesterol-lowering drugs as if that was the magic elixir. Not so simple.

Instead, our skin is getting peeled back for a quick look inside. This is the end of medicine as we know it. Don't guess that I might have hardening of the arteries. Open me up and take a look. Don't guess that I don't have cancer because I'm not spitting up blood or growing a tumor the size of a grapefruit out my side. Quit squinting at X rays of breasts and send in the marines in the form of probes who land on the beach and send up flares if they detect the enemy.

Maybe 90 will be the new 75.

Maybe NBC's *Today Show* will quit celebrating centennial birthdays because they can't fit them all in. Is 100 the new 80?

It's not so much living longer, but being healthy enough to enjoy it. I don't know about you, but I'm not signing up for a life in a hospital bed with tubes flying all over the place and coffee enemas every Thursday. Free as a bird is more like it.

Once imaging becomes mainstream, what kind of bizarre world will we be living in?

"Come on in the family room and see a video of your uncle's colon."

Or do we enter a world of moral hazards. "Philip Morris buys up all the patents for molecular imaging and sells cigarette packs with coupons. Collect 100 coupons and you get a free scan for lung cancer. Smoke until cancer shows up—then we'll clean you up." It's sort of a modern version of smoking through your tracheotomy. Don't knock it until you've tried it.

I think back to R2 Technology's computer-aided detection algorithms. They train the machine by scanning in thousands of mammograms from breasts with cancer. Will we scan cadavers looking for patterns related to death? I hope so.

Do video game companies get involved? Do massively multiplayer games searching for and destroying mutants become massive multipatient systems? Navigate your Uncle Ira and search and destroy malignancies and mutant genes.

Here's one I can't get out of my head in the middle of the night: Do we all keep Tupperware filled with hairless, pink mice? These personal mice become our own factories for just the antibodies to fight what ails us. No more spaghetti against the wall—throw your mouse against the wall and harvest its spleen. Not quite *Planet of the Apes*. Creepy, yet somehow appealing.

The Japanese go overboard, or is it overcommode, with their bathroom tools. Smart toilets that flush, spray, blow hot air and send you on your way are only missing the robotic toilet paper–clutching hand. But why not put Sam's sensor chip in your toidy?

"Good morning, Andy. Your vitamin C is low, I'm getting early indications of pancreatic cancer and you'll never get rid of your tennis elbow if . . ." Okay, thank you.

Implantable chips? I suppose. But the last thing I want is for my body to start producing proteins to fight the antibodies that were produced to fight the chip containing 10,000 antibodies to detect the protein that was just produced. A little circular for my taste. Hurricane producing, as my grandmother might say, since we touched something in there.

There is already a movement to capture the DNA of each of us. I suppose scanning our DNA is easier than scanning cadavers. Craig Venter has offered a prize for anyone coming up with a process to do personal sequencing for under $1,000. No takers so far, but I've got to believe it's just a matter of time.

There is an International HapMap Project, a kind of open-source project to create haplotype maps, patterns of differences of the human genome. If everyone who develops pancreatic cancer has similar genes, actually the same differences in genetic code from the rest of the population, well, knowledge is there. How much? Too early to tell.

For the day when sequencing is really cheap, some folks at Harvard Medical already have announced a Personal Genome Project—an open-source effort along the lines of Linux and software development. Except they have a few concerns for you to think about before you contribute your own DNA sequence. Consider yourself warned:

1 Unwanted paternity accusations
2 A prospective employer may think you are likely to get colon cancer and pass
3 You may actually be related to Lee Harvey Oswald and not have wanted to know this
4 Someone may plant a copy of your DNA at a bank robbery and get you hauled off to the slammer
5 You may actually have a rare disease with no cure and die of anxiety

Is this what they think about at Harvard? Someone has to. They left off the most obvious problem with an open database of DNA. The knock on the door, most likely in the middle of dinner, and the person says, "Hi, I'm you."

•   •   •

And then there's the digital divide. Who can afford this stuff and who can't? You can already see the writing on the wall: Medicare will be means-tested at some point. Meaning if you can afford it, you don't get the handout from the government. I suspect Social Security will go the same way, but that's another story.

The poor in the U.S. will still get the latest and greatest in health care, especially if early detection is a money *saver* over time. The middle class will pay their own bills, but what else is new? No, the divide is elsewhere, between rich and poor nations. We sell drugs for cheap to other countries with socialized medicine because with the R&D done, there is so little cost in making pills—it's all upside. If you reimport too much from Canada, it kind of defeats the purpose. Canadian prices would have to go up, or worse, cut off the supply.

But you can't dump 1,024-slice scanners on Canada or France. Molecular imaging won't be excess you can just send overseas. Even these cheap protein sensor chips from Intel or whoever else won't be sold just for cost in Africa. It's not going to happen.

Fortunately, supply and demand is one of those basic tenets of economics that won't change. If I knew that there was a chip that could save your life by detecting some killer disease ahead of time, and the U.N. wasn't going to airlift it into my village, I might work my ass off to be able to afford it, and have it delivered by those guys in brown shorts. As it gets cheaper, early detection will be a huge worldwide market.

And then there's another weird divide. The bigger problem is not technology advancing, but its rate of acceptance. Not literal acceptance—as we've seen, doctors will use technology if they can get paid for it, but emotional, psychological acceptance.

I sweated through my clothes while waiting for the results

of my heart scan. My hands trembled as I threw fecal occult blood tissues into the toilet. I wasn't sure I really wanted to know.

I cringed at the thought of taking genetic tests. What if I'm genetically prone to fall down stairs or something?

But then, when the tests were over, I was okay with them. Knowing I had mutated genes was a shock, but one that got me fired up to learn more about colon cancer, figure out what to do rather than sit around and pick my toes and worry.

More seriously, if the Big Three are kicked, even reduced, what do we die from? I come back to this question again and again. It's kind of like: When you get to the end of the universe, and you hit a wall, what's on the other side?

Our biology insists that we die eventually. Cells break down. Old age often comes from just a breakdown of the system. Bones get brittle, organs stop working, things fall off in the middle of the night. I'm okay with that. You can't argue with that design. It's more of a question of getting to that point without incident.

People used to die in famines and floods. Technology has resolved these. Not completely, of course, especially in the non-industrial world, but modern farming is a lot more efficient. And the 100-year-flood or killer hurricane hits only every century or so, as we saw in New Orleans.

Or people die in wars. World War II ended only 60 years ago, so it's hard to make the statement that big world wars are a thing of the past. Satellites and global positioning systems mean you can fight wars with brains versus brawn, an interesting change. Plus, fighting over patents is different from fighting over coal or oil. It's hard to imagine another 60-million-casualty conflict. At least I hope I don't have to.

Then there is the occasional flu pandemic. Those have wiped out quite a few folks, but technology to quickly identify viruses and come up with a workable vaccine seems to be in place.

So it's been heart, stroke, cancer. But now what? Do we all have to be hit by a bus?

Will Dr. Kevorkian personally push us off a cliff when we're done? Scary if that guy lives forever.

But really, where does this all go?

The most likely answer is that even if we reduce deaths from the Big Three, something else will take its place. Obesity rates suggest we are eating ourselves to death. Diabetes isn't going away, despite advances in treatment.

Alzheimer's disease is crippling seemingly healthy seniors. They've beaten heart and stroke and cancer only to see their brains go. 70 is the new 5. Incredibly sad, but hopefully something that scanning and early detection can help to change.

And then there's AIDS and HIV, which came out of nowhere and has inflicted huge swaths of populations worldwide.

Is there something brewing at the bottom of the ocean that will make its way ashore and kick our asses? Good-bye, Columbus?

If the Big Three turns into the Big One, I just hope there's enough time to empty the wine cellar and finally load up all those Marx Brothers DVDs I've been waiting for a rainy day to watch.

Is medicine over? It's all on the margin. I can't visualize the days of boarded-up hospitals. There is always something that comes up. But—and I learned this from my days as an investor, you always say "But"—what if this scale really does start to play out? Does spending peak? Doubtful. It's more likely that the overall health care spending growth rate begins to slow, and money gets spent on different things. We probably won't need as many hospitals, certainly not chronic care hospitals. That includes a huge supply chain of supplies and support, from bedpans to nurses—all this stuff would no longer be a growth business. Maybe there won't be as many ineffective cancer drugs if they consolidate to just a few that work—and work very specifically on cancers that aren't detectable early or treatable by radiation or microsurgery. One thing looks clear—some form of radiology becomes a much bigger field. And I suspect that surgery continues its decline.

It's the flip side that's most interesting. We're gonna need lots more nursing homes. Or extra bedrooms in our kids' homes. Homes on golf courses go on a tear. Cruise ships out of Florida and San Diego see a resurgence. Online bridge replaces poker as the new hot fad.

I can see Social Security and Medicare being means tested. Dean Kamen needs to dust off the plans for his Segway personal transportation device and make it senior-friendly. Bland food makes a comeback. The Rolling Stones set up permanently at Caesars in Las Vegas. A pretty scary thought.

And me? Do I think I'm going to live forever? Our economy and our technology can and will do spectacular things. I've seen it, invested in it, predicted it, been wrong, been right. I don't see a singularity—all of us living to 130 and backing up our brains and regrowing organs. As my grandmother might have said, there are too many hurricanes inside of us. But I do see the definition of old getting pushed further and further out. 70 is no longer old. Maybe soon, 100 won't be old. And then I think it's the same for everyone in my generation. I hope I die before I get old. But just before.

And as Coach Dan Reeves told Dr. John Simpson: Even if you are winning, cheat 'em a little bit anyway.

Chapter 59

∎

Singing Canary

I drove past the turnoff for Bracewell's satellite dishes on my way up the hill and smiled. Most of Portola Valley is tucked into the folds of a set of mountains that separate the San Francisco Bay and Silicon Valley from the wilds of the Pacific Ocean. Every night the fog rolls in, and then almost every morning, although sometimes it takes until lunch, the fog slowly rolls back out as the sun heats up the air. I kept driving and driving, around winding curves and turnoffs, until finally I hit the dead end, okay, cul-de-sac, and the house at the end of the road.

I was invited to attend a Cancer Luncheon at Sam Gambhir's house. I wasn't sure what to expect—patients in bandannas, schmattes, do-rags, fresh from Gilda's Club recovering from chemo? Or perhaps researchers arguing about care versus early detection? I had no idea.

Instead, I walked into a room filled with extremely well-dressed women—lots of them, milling around, chitchatting, a coffee klatch deluxe. I recognized a few of them, and practiced my perfected art of reading name tags without being noticed on the rest. This was no high tea or Bunco tournament; there were several billion dollars' of net worth scattered through Sam

Gambhir's living room. I think an entire inventory at Nordstrom's was emptied that morning preparing for this lunch.

I tried to blend in, but that was a tad difficult, as the only testosterone in the room were me, Sam and Don Listwin. At least I was wearing a light blue shirt.

The conversation slowly faded and Sam's wife, Aruna, got up and introduced the luncheon.

"Many of you may not know, but I was diagnosed with breast cancer eight years ago. But I'm one of the lucky ones, I survived. I can't tell you how much it means to me to see the work that Don Listwin is doing, and that my husband is doing, in early detection so others won't have to go through what I've gone through."

I looked at Sam, who twitched a little in his chair. Sometimes work does touch close to home.

Don got up and started in on his presentation. "Thanks for coming. I just want to fill you in on what we're doing at the Canary Fund. It's quite exciting."

He clicked to slide two, which he looked at, and then continued talking. "My mum died four years ago of . . ." He stopped mid-sentence. The slide had a picture of his mom and Don was misty-eyed. So were half the woman in the room—and I had to admit to personal moisture building. I looked closely at Don. This was no act. I always wonder what motivates those who have enough money to do what they want—and I think I finally figured out what drives Don. It's from pretty deep inside.

". . . ovarian cancer. She was asymptomatic. They thought she had a yeast infection." The slide changed. "You find this stuff pre-Stage I, and there's a 9 in 10 chance of survival. You find it late, and the odds flip—1 in 10 chance of survival. My mum was diagnosed late. They even thought they caught it until the surgeon found more cancer than they had originally thought, spread further then they thought, and . . .

"It's got to change. That's what the Canary Fund is trying to do, turn the odds around—make sure it's 9 out of 10."

Don went on about proteins and biomarkers and the three I's: identify, isolate and intervene. There wasn't a peep in the room.

"It's team science—I've gathered folks at the Hutch in Seattle, at Stanford, like Sam, who will speak in a moment, UCSF, USC. We'll add more, too. Whatever it takes. You'd be surprised at how little is spent on early detection. What we spend actually makes a difference. I like to look at it as a Virtual Institute. Why build a nice building and sell naming rights to wings and floors and labs? Instead, I want to be aggressive on how money is spent, funding the best in other institutions. Our goal is a mass market panel for screening for ovarian cancer and lots of other kinds within a few years. We're almost there."

He blasted through more slides.

"It's just like Cisco. I've got a five-phase plan—I know how to motivate people. We'll get serum markers and imaging to become mainstream. I've got a group creating databases to become the Linux of biotech." I noticed lots of heads nodding. I wouldn't have been surprised if there was a bunch of programmers in the room.

"We need your help," Don concluded. "The plan is to raise 10 times what we've invested so far. I really think that when we show this model works, we'll attract a lot more money in as well, from the National Cancer Institute, and from VCs, and our generation will be the first to benefit."

I didn't get a sense that Don wanted to be Max Bialystock, shaking down well-dressed socialites, but instead that he was trying to get their purses interested in their own future.

With matching gifts, $20 million can be leveraged to $60 million. I wonder what Don Listwin could do with $250 million? How many mass specs can be bought? How many more mice can be imaged? Human experiments are going to be expensive. Or maybe Don's goal of getting those greedy, er, return-seeking venture capitalists involved is closer to reality. Billions are just sitting in bank vaults waiting for something that reaches markets of millions or billions of patients.

Five grand is spent on health care per person every year in the U.S. It's not hard to imagine early detection, imaging, chips and a new style of personalized medicine making up half or more of that in the next 20 years. That's Scale with a capital S and Return on capital with a capital R. Perhaps the Don List-wins of the world won't have to fund this much longer—but instead have played their role of whacking the thick brush out of the way for others.

Maybe the early detection brakes aren't locking up on the active runway of runaway spending on chronic patients. It feels like it's not the end of road in Portola Valley, at Stanford, at the Hutch, or for scanning. Instead, it feels like the start of something big—probably just the end of medicine as we know it.

Acknowledgments

Heading into the unknown of medicine (unknown to me anyway) required many guides, most of whom are on the pages of this book. But I especially want note my gratitude to both Dr. Gary Glazer and Dr. Sam Gambhir at Stanford for teaching me what is possible, and opening up their world for me to observe. I also want to thank Don Listwin for his help and access, and vision of where technology and medicine cross paths. And when I got lost, venture capitalist Jon Medved pointed me back in the right direction.

A few folks had the courage to read through early, typo-ridden versions of the manuscript and help mold it into something useful. I want to thank Amy Finnerty, Dr. William Forman (also known as my cousin Bill), and Derek Lowe of the informative blog pipeline.corante.com.

Thanks to my agent Daniel Greenberg for shaping my early thoughts.

The team at HarperCollins deserves lots of thanks, especially Joe Tessitore for believing in the project. Thanks also to Beth Mellow, Larry Hughes, and Libby Jordan for getting the word out and to Alex Scordelis for getting the words in the right form.

This book would never have happened without the vision

and prodding of my publisher Marion Maneker. I proposed several versions of this book, with no response, until one day Marion called up and said, "So what's taking you so long with this book." That's all the encouragement I needed. I am always looking for useful advice, and remember fondly Marion's comment, "You know, this would be a lot easier book to write if you just had a heart attack." Well, fortunately, I'm not one to take the easy way out. Sorry to disappoint, but a sincere thanks for everything you have done in teaching me how to turn an idea into a fun and readable book.

And finally, I'd like to thank my wife, Nancy, for being the best partner anyone could have and allowing me to feature her in a drooling, mouse-fearing cameo role. Your love is the only medicine I need.

Index